The IMA Volumes
in Mathematics
and its Applications

Volume 37

Series Editors
Avner Friedman   Willard Miller, Jr.

Institute for Mathematics and
its Applications
IMA

The **Institute for Mathematics and its Applications** was established by a grant from the National Science Foundation to the University of Minnesota in 1982. The IMA seeks to encourage the development and study of fresh mathematical concepts and questions of concern to the other sciences by bringing together mathematicians and scientists from diverse fields in an atmosphere that will stimulate discussion and collaboration.

The IMA Volumes are intended to involve the broader scientific community in this process.

Avner Friedman, Director
Willard Miller, Jr., Associate Director

* * * * * * * * * *

IMA PROGRAMS

* * * * * * * * * *

SPRINGER LECTURE NOTES FROM THE IMA:

Rutherford Aris   Donald G. Aronson
Harry L. Swinney
Editors

Patterns and Dynamics in Reactive Media

With 64 Illustrations

Springer-Verlag
New York Berlin Heidelberg London Paris
Tokyo Hong Kong Barcelona Budapest

Rutherford Aris
Department of Chemical Engineering
and Material Science
University of Minnesota
Minneapolis, MN 55455
USA

Donald G. Aronson
School of Mathematics
University of Minnesota
Minneapolis, MN 55455
USA

Harry L. Swinney
Physics Department and the
Center for Nonlinear Dynamics
University of Texas
Austin, TX 78712
USA

Series Editors
Avner Friedman
Willard Miller, Jr.
Institute for Mathematics
and its Applications
University of Minnesota
Minneapolis, MN 55455
USA

Mathematics Subject Classifications: 80A30, 80A32

Library of Congress Cataloging-in-Publication Data
Patterns and dynamics in reactive media / Rutherford Aris, Donald G.
   Aronson, Harry L. Swinney, editors.
      p.   cm. — (the IMA volumes in mathematics and its
   applications ; v. 37)
      Contains lectures given at a workshop held at the Institute for
   Mathematics and its Applications, Minneapolis, Minn., Oct. 16–20,
   1989.
      Includes bibliographical references.
      ISBN 0-387-97671-X (N.Y. : alk. paper). — ISBN 3-540-97671-X
   (Berlin : alk. paper)
      1. Fluid dynamics — Congresses.   2. Combustion — Mathematical
   models — Congresses.   I. Aris, Rutherford.   II. Aronson, Donald G.
   III. Swinney, H. L., 1939–   .   IV. Series.
   QA911.P38   1991
   541.3′61 — dc20   91-28048

Printed on acid-free paper.

Camera-ready copy prepared by the IMA.
Printed and bound by Edwards Brothers, Inc., Ann Arbor, MI.
Printed in the United States of America.

9 8 7 6 5 4 3 2 1

ISBN 0-387-97671-X Springer-Verlag New York Berlin Heidelberg
ISBN 3-540-97671-X Springer-Verlag Berlin Heidelberg New York

mathematics

The IMA Volumes
in Mathematics and its Applications

Current Volumes:

FOREWORD

This IMA Volume in Mathematics and its Applications

PATTERNS AND DYNAMICS IN REACTIVE MEDIA

is based on the proceedings of a workshop which was an integral part of the 1989-90 IMA program on "Dynamical Systems and their Applications". The aim of this workshop was to cross fertilize research groups working in topics of current interest in combustion dynamics and mathematical methods applicable thereto. We thank Shui-Nee Chow, Martin Golubitsky, Richard McGehee, George R. Sell, Rutherford Aris, Donald Aronson, Paul Fife and Harry Swinney for organizing the meetings. We especially thank Harry Swinney, Rutherford Aris and Donald Aronson for editing the proceedings.

We also take this opportunity to thank those agencies whose financial support made the workshop possible: the Army Research Office, the National Science Foundation and the Office of Naval Research.

Avner Friedman

Willard Miller, Jr.

PREFACE

This volume contains some of the lectures given at the workshop "Patterns and Dynamics in Reactive Media" held from October 16–20, 1989 as part of the year on Dynamical Systems and their Applications at the Institute for Mathematics and its Applications, Minneapolis, Minnesota.

Ever since the seminal works on traveling waves and on morphogenesis by Fisher, by Kolmogorov, Petrovski & Piscunov, and by Turing, scientists from many disciplines have been fascinated by questions concerning the formation of steady or dynamic patterns in reactive media. The contributors to this volume include chemists, chemical engineers, mathematicians (both pure and applied), and physicists. Their contributions range from reports of experimental studies, through descriptions of numerical experiments, to rather abstract theoretical investigations, each exhibiting different aspects of a very diverse field. Although this small volume can hardly claim to cover the whole range of current research in patterns in reactive media, it nevertheless presents a representative sample.

It is a pleasure to thank all of the lecturers and participants for a very stimulating workshop. We are also especially grateful to the staff and directors of the IMA for courteously and efficiently provididng a very rich medium in which we could all react (and form new patterns) . Finally, Patricia Brick, Stephan Skogerboe, and Kaye Smith deserve both thanks and a medal for their patient work in producing this volume.

Rutherford Aris

Donald G. Aronson

Harry L. Swinney

CONTENTS

SIMPLE RESONANCE REGIONS OF TORUS DIFFEOMORPHISMS

CLAUDE BAESENS†, JOHN GUCKENHEIMER‡,
SEUNGHWAN KIM§ AND ROBERT MACKAY†

Abstract. This paper discusses resonance regions for two parameter families of diffeomorphisms and vector fields on the two dimensional torus. Resonance regions with at most two resonant periodic orbits (in the discrete case) or two equilibrium points (for flows) are studied. We establish global geometric properties of these regions with topological arguments.

Key words. dynamical systems, bifurcation theory

AMS(MOS) subject classifications.

This paper discusses the geometry of the simplest resonance regions displayed by three weakly coupled oscillators. Much more extensive descriptions of the dynamics of three coupled oscillators are given in Baesens, Guckenheimer, Kim and Mackay [1]. Indeed, the results presented here are a footnote to [1] and the reader should look to this work for diagrams that illustrate the phenomena discussed here. Generic two parameter families of three weakly coupled oscillators give vector fields in which there is a smoothly varying family of invariant three dimensional tori. By taking cross-sections of the flows on the invariant three dimensional tori, the three interacting frequencies can be studied via a family of diffeomorphisms of the two torus. This is the setting in which we shall work.

The theory of diffeomorphisms of the circle is thoroughly developed. It yields information about resonances and mode lockings of two interacting frequencies. There are stringent limits on the dynamical possibilities. The dynamical behavior of a circle diffeomorphism can be summarized via its rotation number ρ. If ρ is irrational, then the motion is quasiperiodic. Otherwise the rotation number is rational, and all trajectories tend to periodic orbits of the same period in both forwards and backwards time.

The theory of diffeomorphisms of the two torus is much more complicated than the theory of diffeomorphisms of the circle. Phase portraits on a two dimensional torus can have multiple attractors as well as chaotic trajectories. Forward and backward limit sets of different types of different types can coexist. The three types of limit sets that are seen most frequently in generic diffeomorphisms are periodic orbits, invariant curves and the whole two torus. Examples with chaotic trajectories occur, but they are seldom found without a systematic search. A natural setting for the study of diffeomorphisms of the two torus is the investigation of families in which the amount of rotation in the two directions around the torus can be varied. The family of diffeomorphisms $f_\Omega(x) = x + \Omega$ is called the family of **translations**.

†Mathematics Institute, University of Warwick, Coventry, England CV47AL

‡Department of Mathematics, Cornell University, Ithaca, NY 14853. The work of the first author was supported by the National Science Foundation and the Air Force Office of Scientific Research.

§Center for Applied Mathematics, Cornell University, Ithaca, NY 14853

Here the two dimensional parameter Ω varies the amount of rotation. Most of the theory that has been developed thus far has been within the context of families of the form $f_\Omega(x) = x + \Omega + \epsilon g(x)$ with $g(x)$ a nonlinear, doubly periodic function and ϵ a third parameter which controls the size of nonlinearity. For fixed ϵ, the system goes through regions with many different rotation vectors as Ω is varied.

One can begin to understand the dynamics of a generic family of diffeomorphisms of the two torus by first considering the family of rotations and then perturbations of this family. In a generic two-parameter family that is close to the family of rotations, KAM theory guarantees that the sets of parameters values for which the corresponding diffeomorphisms are not smoothly equivalent to irrational rotations occupies a small measure in the parameter space. The diffeomorphisms with limit sets that are periodic orbits or invariant curves are the object of our interest. We recall some properties of the parameter sets giving rise to this mode-locked behavior and of the dynamics that are observed on the torus when the mode-locking occurs.

In the product of the parameter and the phase spaces, there is a two torus of periodic orbits corresponding to each rational parameter vector $(p_1/q_1, p_2/q_2)$. The period q of these orbits is the least common multiple of q_1 and q_2. It is easy to check that this torus is a nonsingular set of roots of the pair of equations $f_\Omega^q(x) = x$ on the product space. It follows that a family of torus diffeomorphisms that are near rotations will have a torus P of periodic orbits of period q. Unlike the family of translations, one cannot expect that P will project to a single point in the parameter space. The image of the projection is called the **resonance region** of the rotation vector with $(p_1/q_1, p_2/q_2)$ associated with P. Basic properties of resonance regions are described in [3]. As the image of the map of a torus into the plane, there are topological constraints on the form of the resonance region and its boundary [4]. In particular, generic families have resonance regions whose only singularities are cusp points. Moreover, a resonance region cannot have a connected boundary without cusp points. The presence of cusp points implies there are parameter regions in which there are four periodic orbits with rotation vector $(p_1/q_1, p_2/q_2)$.

There are resonance regions of families of diffeomorphisms for which the individual systems contain only one or two periodic orbits of the resonance rotation vector. These resonance regions appear to arise naturally in two ways. First, if one starts with a "Mathieu" family in which the nonlinear part $\epsilon g(x)$ of the family $f_\Omega(x) = x + \Omega + \epsilon g(x)$ is given by a pair of trigonometric monomials, then the expansion of $f^n(x) - x$ near a resonance region of order n as a power series in ϵ will normally have dominant and subdominant terms that are trigonometric monomials. Galkin [2] describes the combinatorics associated with determining which are the dominant terms and their degrees. The new results of this paper address the geometry of these "simplest" resonance regions. The global topology of the torus plays a prominent role in the discussion.

The behavior associated with invariant curves is more complicated. For each rotation vector $\Omega = (\omega_1, \omega_2)$ satisfying only one independent equation $p_1\omega_1 + p_2\omega_2 = q$ with integer coefficients, there is also a two dimensional torus in the product of the phase space and the parameter space for the family of translations consisting of orbits that have this rotation vector. The trajectories on this torus are quasiperiodic, but lie on closed curves whose homotopy type is determined by Ω. For perturbations, these tori may or not persist: KAM theory provides the techniques proving persistence for "good" rotation vectors. As Ω varies over vectors satisfying the relation $p_1\omega_1 + p_2\omega_2 = q$, one obtains a line in the parameter space for the family of translations along which there are invariant curves of the same homotopy type for the corresponding diffeomorphisms. For perturbations of the family of translations, the first approximation to the set of parameters with invariant curves of the specified homotopy type is a strip. As one moves along this strip, there are many points at which the corresponding diffeomorphism acts like an irrational rotation on a smooth invariant curve. At other points of the parameter space, there may be invariant curves with less smoothness that are formed from saddle separatrices. There may also be holes in the strips in which there is no closed invariant curve at all. In addition to the **rotational** invariant curves that have non-trivial homotopy type, one encounters invariant curves that can be continuously deformed to a point on the torus.

The basic framework of a set of overlapping resonance regions and strips associated with invariant curves of specific homotopy types can be substantially refined by consideration of bifurcations showing the behavior at the boundaries of the strips and resonance regions and the transitions that take place inside them. The geometry associated with these bifurcations is complex, but there are several basic facts and principles that can be used to understand this geometry.

First, one has the simple fact that two closed curves of different non-trivial homotopy types on the two dimensional torus must intersect. This implies that if rotational invariant curves of different homotopy types exist, they are singular curves that contain periodic orbits. Consequently, strips associated with different homotopy types intersect only in resonance regions.

Second, there is an empirical observation that phase portraits of torus diffeomorphisms near translations usually look similar to those that come from the time one map of flows. This observation is supported by the fact that every torus diffeomorphism has finite Taylor expansions that occur as the Taylor expansions of time 1 maps of flows. Since flows on two dimensional manifolds do not have chaotic trajectories, examining the dynamics of families of time one maps of flows gives us a gentler approach to the study of resonance phenomena for diffeomorphisms. Nonetheless, many of the topological features associated with diffeomorphisms cannot occur in time one maps of flows, so additional analysis is required once one has a good intuition for the dynamics of families of flows.

Third, there is a list of codimension one and two bifurcations that occur in generic two parameter families of two dimensional flows. This list can be used to interpret and guide numerical explorations that seek to delineate the dynamics that occur in specific examples. The interplay between the theoretical analysis of normal forms and interactive computations of phase portraits has played an important role in our developing understanding of families of torus maps. The novel aspect in the analysis of normal forms for our studies has been the necessity of looking carefully at more complicated patterns of saddle connections than have been examined previously.

Recall some of the different types of codimension one and two bifurcations for flows on two dimensional manifolds. In codimension one, there are saddle-node bifurcations at which a pair of equilibrium points coalesce and Hopf bifurcations at which a family of limit cycles ends at an equilibrium. Pairs of limit cycles can coallesce in bifurcations called either double limit cycles or saddle-nodes of periodic orbits. The final type of codimension one bifurcation involves a homoclinic or heteroclinic orbit in which there is a trajectory that terminates at a saddle point as time tends to $\pm\infty$. In codimension two, the list of bifurcations is longer. The local bifurcations involving equilibrium points are cusps, degenerate Hopf bifurcations and the Takens-Bogdanov bifurcations at which an equilibrium has a nilpotent linear part. In addition, there are cusps of limit cycles and various types of bifurcations involving homoclinic and heteroclinic trajectories. For understanding the geometry of the simplest resonance regions occurring in two parameter families, the additional bifurcations that occur involve homoclinic orbits to saddle points with eigenvalues of equal magnitude, saddle-node loops in which there is a homoclinic trajectory for a saddle-node, and double saddle-loops in which a saddle point has two distinct homoclinic trajectories.

Consider a two parameter family of flows on the two dimensional torus with the property that each flow contains at most two equilibrium points. We want to prove as many properties about the bifurcation diagram of such a family as possible.

PROPOSITION. *Let $f : T^2 \to R^2$ be a smooth map with generic singularities and the property that, for each $y \in R^2$, $f^{-1}(y)$ is zero, one or two points. Then the image of f is an annulus with smooth boundaries. The singular set of the map consists of two curves that lie in the same non-trivial homotopy class.*

Proof. The generic singularities of maps between two dimensional manifolds are folds, cusps and the transversal intersection of two fold curves. Since cusps have an adjacent region which is triply covered by the map, cusps do not occur. Nor can there be intersections of two fold curves since each fold has an adjacent region for which there are at least two preimages of each image point. Thus, adjacent to a transversal intersection of two folds is a region in which each point has at least four preimages. We conclude that the singular image of the map is a smooth one dimensional manifold consisting entirely of folds. Thus, the singular image divides the plane into a finite number of regions, one of these regions is the regular image of the map, and the regular image is double covered. Since the Poincaré index of the torus is zero, it follows that there are precisely two boundary components to the image. These components divide the torus into homeomorphic regions. Hence the singular set on the torus must consist of two curves that lie in the same non-trivial homotopy class.

THEOREM. *Let $\dot{x} = \Omega + g(x)$ define a two parameter family of flows on the torus with the property that for each Ω there are at most two equilibrium points of the corresponding flow.*

(1) *The resonance region for the map is an annulus.*

(2) *The flows in the bounded component of the complement of the resonance region (the "hole") do not have a cross-section. In the notation of [1], the flow has type D_ω for a fixed rational ω.*

(3) *There are at least two curves of Hopf bifurcations and six curves of homoclinic bifurcations in the resonance region.*

(4) *The curves of homoclinic bifurcations meet in at least two necklace points at which all of the saddle separatrices of the flow are homoclinic.*

Proof. The first statement is a consequence of the previous proposition since the resonance region is the image of the map $-g : T^2 \to R^2$. The existence of at least two Takens-Bogdanov points on each boundary of the resonance region was proved in an earlier paper [3]. This implies the existence of at least two curves of Hopf bifurcations and two curves of saddle-node bifurcations. The points on the torus at which the map $-g : T^2 \to R^2$ has a fold give rise to the saddle-node bifurcations for appropriate parameter values (that depend on the point). Let C be the curve on the torus whose image forms the inner component of the boundary of the image of $-g$. Then C is a smooth closed curve that is homotopically non-trivial on the torus. Observe that if Ω is a parameter value in the hole, then the image of C by the map $-g$ has winding number ± 1 with respect to Ω. Thus, the map $g + \Omega : T^2 \to R^2$ has image that is an annulus, the origin is in the bounded region of its complement and the image of C has winding number ± 1 with respect to the origin. We assert that this is incompatible with the presence of a cross-section to the flow.

If a flow has a cross-section, then the winding number of the vector field with respect to the origin is zero along any closed curve on the torus. This is readily seen by constructing a homotopy of the vector field to a constant vector field through vector fields without fixed points. Winding vectors do not change under such homotopies. The construction of the homotopy can be accomplished by an initial homotopy that makes the vector field have a constant return time. It is then transverse to a one parameter family of cross-sections. A second homotopy of coordinate changes makes these cross-sections parallel to one another. Finally, a homotopy of the vector field makes it of unit length and orthogonal to the parallel cross-sections. It is then constant. We conclude that $g(x) + \Omega$ has no cross-section.

Consider the vector field associated to a parameter value inside the hole of the resonance region. There are at least two periodic orbits with opposite orientations, but we can say more as well. The torus is divided into annuli by the periodic orbits. Each of these annuli has either trajectories whose orientation does not change or trajectories whose orientation changes by $\pm\pi$. The second type of annulus is called a **Reeb component**. The sum of the changes in orientation from all the annuli is $\pm 2\pi$, corresponding to the index change of the vector field as we traverse the curve along which folds occur in the vector field. If we collapse the annuli in which orientation does not change onto closed curves, we obtain a topological flow on a torus with periodic orbits separated by Reeb components. At least one of the periodic orbits must be attracting (and hence one repelling) since otherwise the sum of changes in orientation from the Reeb components would be 0.

The flow of vector fields for parameters in the hole have annuli that are attracting their boundaries contained in the Reeb components. The presence of an attracting annulus with boundary in Reeb components is a persistent phenomenon under perturbation. Though the size of an attracting annulus may change discontinuously as periodic orbits undergo bifurcation, perturbations of the flow will have an attracting annulus that intersects the original one. As our parameter varies over the disk, we can therefore find continuous families of closed curves that lie in the the Reeb components that contain the boundaries of an attracting annulus.

We next consider the geometry of the vector fields obtained by traversing the parameter curve γ bounding the hole. For each parameter value, there is still a periodic orbit of the homotopy type of those found in the hole since a saddle-node bifurcation will not affect periodic orbits of opposite orientations. Observe that the curve σ traversed on the torus by the saddle-node points lies in a different (non-trivial) homotopy type from the periodic orbits since the winding number of the vector field along this curve is non-zero for parameter values inside the hole. We also assert that the saddle-node points cross Reeb components containing the boundary of an attracting annulus as one traverses γ. If this were not the case, then a continuous family of closed curves in the Reeb components would undergo a non-trivial homotopy along σ. This is a contradiction since these families of closed curves in the Reeb components can be extended continuously over the hole, implying that the homotopy is trivial.

Since the saddle-nodes cross Reeb components as parameters traverse γ, there must be parameter values in γ for which the saddle-node points cross the boundaries of Reeb components. In a generic family of vector fields, these must be codimension two bifurcations which adjoin regions with different numbers of periodic orbits. Examining the list of codimension two bifurcations for families of vector fields with at most one saddle point, the only possibility for these bifurcations are that they are **saddle-node loops**. If the saddle-node has a negative eigenvalue, this means that the unstable separatrix of the saddle-node lies on the boundary of the stable manifold of the saddle-node. Pursuing this logic further, one finds that there must be at least four such bifurcation points corresponding to the boundary components of two distinct Reeb components. ¿From each of these codimension two bifurcations emerges a curve of homoclinic bifurcations in the parameter space. These bifurcation curves are distinct from those that end at the Takens-Bogdanov points since the homoclinic orbits are homotopically non-trivial. We conclude that there are at least six curves of homoclinic bifurcations in our family of vector fields. This ends the proof of the theorem.

REMARKS. The homoclinic bifurcation curves can end only at saddle-node loops in generic families with at most one saddle point. The arguments given above can be extended to show that the homoclinic bifurcation curves that we have identified must cross the resonance region from one boundary component to the other. If one tracks periodic orbits along a homotopically non-trivial closed curve in the resonance region, they follow paths that are homotopic to σ. Thus the saddle-ponts cross "Reeb components" as one traverses the parameter curve. The changes of behavior that occur in the position of saddle points correspond to homoclinic bifurcations. Since the stability of the two boundary components of a Reeb component differ, there must be codimension two bifurcations at which the rotational homoclinic orbits come from a trace zero saddle point. Generically, there are curves of saddle-node bifurcations for periodic orbits that emanate from such codimension two bifurcations. Further analysis of the order with which different homoclinic bifurcations will be encountered on the two boundary components of a Reeb component indicates that the homoclinic bifurcations cross. Thus there are codimension two bifurcation points with double homoclinic cycles. These bifurcations are called **necklace** points in [1].

When one considers the simplest resonance regions for diffeomorphisms rather than flows, the structures described above become more complicated. Curves of homoclinic bifurcations become thickened and the structure of saddle-nodes of periodic orbits becomes more complicated, with the addition of "Chenciner bubbles" where a pair of resonant invariant curves try to approach each other. The reader is referred to [1] for a discussion of what we know about the details of these additional complications that appear in the bifurcation diagrams of two parameter families of diffeomorphisms of the two dimensional torus.

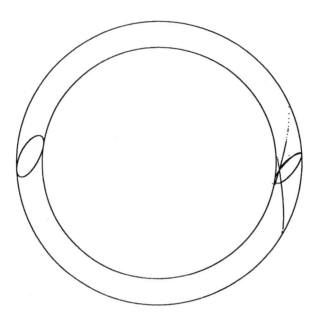

FIGURE 1: A numerically computed simple resonance region for the family of flows

$$\dot{x} = \Omega_x + \cos(2\pi x) + 0.1\cos(2\pi y)$$
$$\dot{y} = \Omega_y + \cos(2\pi x) + 0.1\cos(2\pi y)$$

The entire resonance region with its boundary of saddle-node curves is shown. Half of each small oval is a curve of Hopf bifurcations and half is a curve along which there is a saddle point with determinant one. Curves of parameters with homoclinic bifurcations are shown on one side of the resonance region.

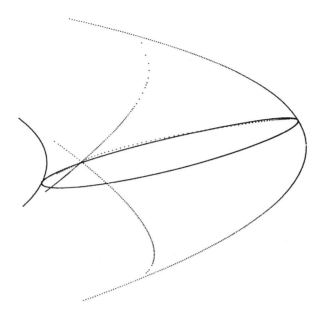

FIGURE 2: An enlarged view of one side of the simple resonance
region shown in Figure 1.

REFERENCES

[1] C. BAESENS, J. GUCKENHEIMER, S. KIM AND R.S. MACKAY, *Three Coupled Oscillators:*
     Mode-locking, Global Bifurcations and Toroidal Chaos, preprint, (), pp..
[2] O.G. GALKIN, *Resonance regions for Mathieu type dynamical systems on a torus*, Physica,
     39D (1989), pp. 287–298.
[3] S. KIM, R.S. MACKAY AND J. GUCKENHEIMER, *Resonance regions for families of torus maps*,
     Nonlinearity, 2 (1989), pp. 391–404.
[4] H. LEVINE, *Stable maps: an introduction with low dimensional examples*, Bol. Soc. Bras.
     Mat., 7 (1976), pp. 145–84.

A MINIMAL MODEL FOR SPATIO–TEMPORAL PATTERNS IN THIN FILM FLOW

H.S. BROWN*, I.G. KEVREKIDIS * AND M.S. JOLLY†

Abstract. We consider the development of spatio–temporal oscillations in the Kuramoto–Sivashinsky amplitude model of thin film flow. These develop from Hopf bifurcations off of steady state solutions and are observed to undergo symmetry breaking and period doubling bifurcations. Oscillatory branches in the parameter regime studied apparently terminate in Silnikov type homoclinic connections. A minimal, three mode nonlinear Galerkin discretization, capable of capturing this bifurcation behavior is constructed. A simple shooting algorithm which exploits this sharp reduction in dimensionality is used to accurately locate the homoclinic connections.

Key words: Kuramoto–Sivashinsky equation, Silnikov connections, symmetry breaking, inertial manifolds.

1. Introduction. The Kuramoto–Sivashinsky equation (KSE) [27, 36]

$$(1) \qquad v_t + 4v_{xxxx} + \alpha \left(v v_x + v_{xx} \right) = 0.$$

has been used as an amplitude equation to describe incipient instabilities in several physical contexts. It can be derived from a long–wave perturbation analysis of the two dimensional Navier–Stokes Equations describing the flow of a thin, viscous liquid film with surface tension, flowing down a vertical plane. The equation has been the subject of numerous theoretical and computational studies during the last fifteen years (e.g. [21, 9, 26, 19] and references therein) probably because of its ability to exhibit low–dimensional, spatially coherent but temporally complicated solution patterns. Numerical simulations have uncovered a wealth of such patterns, ranging from steady states, periodic oscillations and traveling waves to modulated traveling waves, symmetry–driven persistent homoclinic loops, and complicated (apparently chaotic) dynamics as the instability parameter α grows. Theoretical work based largely on symmetry considerations has concentrated on steady state and traveling wave solutions, as well as heteroclinic and homoclinic connections between them (e.g. [1]).

In this work we concentrate on the computer assisted study of limit cycle (time periodic) solution branches arising from low α Hopf bifurcations. Two types of periodic solutions are observed, one possessing a spatio–temporal symmetry that relates the first half of the period of the oscillation with the second half. Such a symmetry is known to suppress period doubling of the oscillations [38]. A symmetry breaking pitchfork bifurcation follows, with subsequent period doublings of the non-symmetric limit cycles. All oscillatory branches apparently terminate on Silnikov type infinite period homoclinic connections (heteroclinic connections for the symmetric limit cycles), involving saddle type steady state solutions. While Silnikov saddle connections can occur in a phase space of dimension three, the simple three mode Galerkin truncation of the PDE (1) cannot capture these phenomena. However, using the theory

* Department of Chemical Engineering, Princeton University Princeton, NJ 08544

† Department of Mathematics, Indiana University Bloomington, IN 47405

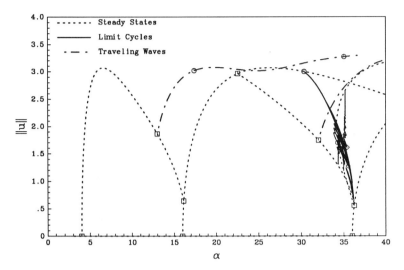

Figure 1: Bifurcation diagram for the KSE; □ marks steady state and traveling wave bifurcations; o marks Hopf bifurcation points.

of approximate inertial manifolds, we construct a minimal (three dimensional) *nonlinear* Galerkin discretization which qualitatively (and to some extent quantitatively) captures the Silnikov behavior in the parameter regime studied. This sharp reduction in the dimensionality of the problem allows for the easy visualization of the stable and unstable manifolds involved in these interactions. We have exploited this to implement a simple shooting algorithm which converges on the Silnikov loop in (phase × parameter) space.

2. Numerical Bifurcation Diagrams. Figure 1 is a bifurcation diagram for the KSE for $0 < \alpha < 40$. The diagram shows steady, periodic, and traveling wave solution branches computed using a traditional Galerkin spectral discretization of the PDE with nine Fourier modes. The diagram is "converged" in the sense that doubling the dimension of the discretization does not visibly alter the location or the stability of the solution branches.

Our calculations were actually performed using the integrated form of the KSE with periodic boundary conditions on $0 < x < 2\pi$:

$$(2) \qquad u_t + 4u_{xxxx} + \alpha \left(\frac{1}{2}(u_x)^2 + u_{xx} - \frac{1}{4\pi} \int_0^{2\pi} (u_x)^2 \, dx \right) = 0.$$

AUTO, a continuation/bifurcation package developed by E. Doedel [12, 13] was used to perform these calculations. The steady state and limit cycle branches (the latter are often called "standing waves") were calculated by restricting the dynamics to the invariant subspace of even functions; all subsequent calculations in this paper will be similarly restricted. Until recently most of the rigorous results (dissipativity, existence and dimension estimates for inertial manifolds) for the KSE have been limited to this invariant subspace (see for example [33]). Il'yashenko [22] has recently presented a proof of dissipativity for the general case of periodic boundary conditions. The traveling wave branches, computed in full Fourier space with periodic boundary conditions, are included here for completeness. Another form of solution (modulated traveling waves) also known to exist in the full Fourier space within this parameter range have not been included in the diagram, since the discussion will be mostly restricted to even solutions and their bifurcations. Such bifurcation diagrams, particularly those including the steady state solutions of the KSE have appeared in the literature (e.g. [34, 19, 3]). In this paper we focus on the family of limit cycle branches observed for $30 < \alpha < 37$. Certain aspects of these branches and their bifurcations have been discussed previously in [24, 23, 7].

The steady state branches bifurcating at $\alpha = 4(= 4 * 1^2), 16(= 4 * 2^2), 36(= 4 * 3^2), \ldots 4 * k^2$ are characterized as "unimodal" $(k = 1)$, "bimodal" $(k = 2)$, ... "k-modal" respectively, based on their spatial structure close to the respective bifurcations. A Hopf bifurcation occurs at $\alpha \sim 30.34$ off of the bimodal steady state branch (restricted on the invariant subspace of even functions). This limit cycle with period T (i.e. $u(x, t) = u(x, t + T)$) is characterized by a spatio–temporal symmetry: the second half period of the oscillation (in time) is related to the first half period by:

$$u(x, t) = u(x + \pi, t + T/2).$$

This symmetry is a version of what Aronson *et al.* call "Ponies On a Merry-go-round" for discrete arrays of Josephson junctions [2]. The implications of such a symmetry on the bifurcation behavior of periodic solutions was discussed by Swift and Wiesenfeld [38].

To understand the properties of these symmetric limit cycles, it is helpful to consider certain replication and invariance properties of the solutions of the KSE. It is easy to see that the space

$$C_k \equiv \mathrm{span}(\cos(kx), \cos(2kx), \cos(3kx) \ldots)$$

is invariant under the flow of the KSE and that if $u(x, t)$ is a solution of the KSE for some parameter value α_0, then so is $u(kx, k^4 t)$ at $\alpha = k^2 \alpha_0$. This means that steady state branches of the KSE replicate as discussed in detail in [34]. For every nontrivial steady state branch, a one–parameter family of steady states exists at every α value, each one of them a shift (in x) of the others. All steady state branches shown in figure 1 (unimodal, bimodal, trimodal, as well as a mixed mode branch, the "bi–tri"), represent solutions which can be shifted to be even (i.e. can be represented by a cosine series on $[0, 2\pi]$). Applying the above replication property with $k = -1$, we see that two "representatives" of each nontrivial k–modal steady state branch can be found in C_k, related to each other by a translation by π/k. The bi–tri steady state branch can be thought of as "unimodal" in this context: two representatives of it exist in C_1 related to each other by a translation by $\pi/1$.

We are interested here in the bifurcations of these even solutions, taking place in the invariant subspace of even functions. The two representatives of odd–modal $(k = 1, 3, 5, \ldots)$ branches can be shown to have the same stability in C_1; on the other hand, the two representatives of even–modal $(k = 2, 4, 6, \ldots)$ branches have different stability properties in C_1 (while of course they have the same stability in C_k as well as in the full Fourier space, where they also both possess a zero eigenvalue corresponding to the direction of shift invariance). The Hopf bifurcation observed at $\alpha \sim 30.34$ is

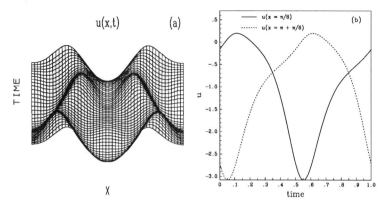

Figure 2: (a) A space–time plot of one period of a symmetric limit cycle ($0 < x < 2\pi$) at $\alpha \sim 33.56$. (b) Time series of the amplitude at $x = \pi/8$ and $x = \pi + \pi/8$ during one period of the oscillation.

not a replication of a lower α Hopf bifurcation on the unimodal branch. Since the eigenvalues (eigenvectors) of the bimodal representatives in C_2 are identical (replicas) to those of the unimodal representatives in C_1, this Hopf bifurcation does not occur entirely *within* C_2 and hence involves components in C_1, in which the stability of the two bimodal representatives is different. The Hopf bifurcation is therefore expected to (and indeed observed to) occur for only one of the two representatives of the bimodal branch.

We now restrict our analysis to the invariant subspace of even functions C_1. The two bimodal representatives are each invariant under π shifts and transformed into each other by $\pi/2$ shifts. Since the replication law with $k = -1$ (the shift by π) is a property of both steady and time dependent solutions, the unique limit cycle resulting from the single representative Hopf bifurcation must be (as a solution) invariant to a spatial shift by π. In this case, the spatio–temporal symmetry:

$$u(x) \to S_\pi u(x); \quad t \to t + T/2$$

results, where

$$S_\pi u(x) \equiv u(x + \pi).$$

The symmetry of the limit cycles is illustrated in figure 2. To see how this symmetry will appear in Fourier space, consider the ODE system

$$\dot{a}_n = F_n(a); \quad n = 1, \ldots, N$$

arising from the Fourier–Galerkin discretization of the KSE:

$$u(x, t) \sim \sum_{n=1}^{N} (u(x, t), \cos(nx)) \cos(nx) \equiv \sum_{n=1}^{N} a_n(t) \cos(nx).$$

The notation $(\cdot, \cdot)$ refers to the standard inner product on L^2. Figure 3 shows a representative phase portrait and time series characteristic of the symmetric oscillations. This type of symmetry in limit cycles is known to cause suppression of period dou-

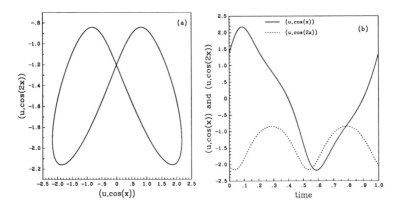

Figure 3: (a) An a_1–a_2 phase space projection of a symmetric limit cycle at $\alpha \sim 33.56$ and (b) time series of two of its Fourier coefficients (a_1 and a_2).

bling. The proof for the KSE follows directly the arguments of Swift and Wiesenfeld, substituting $u(x)$ for their x and S_π (the π shift) for their $-I$. It is also easy to see that the argument carries over for the ODE system resulting from the Galerkin spectral discretization of the PDE. The symmetry becomes

$$
\begin{bmatrix} a_1 \\ a_2 \\ a_3 \\ a_4 \\ \cdot \\ \cdot \end{bmatrix} \rightarrow \begin{bmatrix} -a_1 \\ a_2 \\ -a_3 \\ a_4 \\ \cdot \\ \cdot \end{bmatrix} , \quad t \to t + \frac{T}{2} .
$$

In this case, instead of the Swift and Wiesenfeld matrix $-I$, the diagonal matrix J with elements $j_{ii} = (-1)^i$ is used. The basic argument is that for symmetric orbits, the state transition matrix DP is the square of another matrix $(D\hat{P})^2$ and therefore has as eigenvalues the squares of the eigenvalues of the matrix $D\hat{P}$. The matrices being real precludes DP from having a single real eigenvalue crossing the unit circle at -1 as α varies.

In the KSE (as well as in other systems with this property) a symmetry breaking (pitchfork) bifurcation occurs off of the symmetric limit cycle branch at $\alpha \sim 32.85$, giving rise to two "asymmetric" limit cycle branches (symmetric *to each other* by the π shift in x). These are illustrated in figure 4. The asymmetric limit cycles are not prevented from period doubling, and are indeed observed to subsequently undergo a period doubling cascade [23]. The first period doubling occurs at $\alpha \sim 32.97$ and the second at $\alpha \sim 32.99$. Figure 5 shows representative phase portraits of limit cycles on these branches.

These symmetry properties play a significant role in "orchestrating" the complicated structure of limit cycle branches and associated Silnikov connections shown enlarged in figure 6. Partial observations of this intricate pattern have been presented and published previously in [25, 24, 23, 7, 8]; in what follows we attempt a more coherent discussion of the detailed bifurcation picture. Figure 7a shows the fate

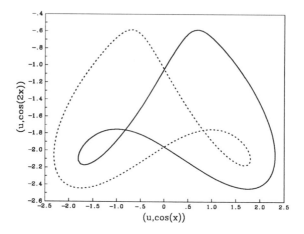

Figure 4: An a_1–a_2 phase space projection of an asymmetric limit cycle and its π shift at $\alpha \sim 33.56$.

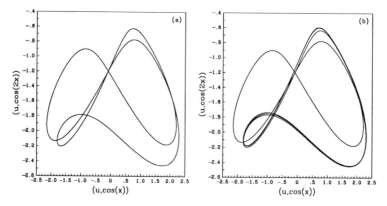

Figure 5: a_1–a_2 phase space projections of (a) the period 2 and (b) the period 4 asymmetric limit cycles at $\alpha \sim 33.56$.

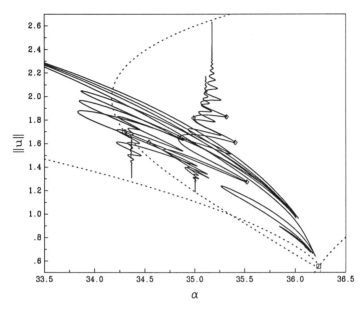

Figure 6: Enlarged bifurcation diagram for the KSE. □ marks steady state bifurcations, o marks Hopf bifurcation points, and ◇ marks period doubling bifurcations.

of both asymmetric limit cycle branches; after the first period doubling, the asymmetric branches undergo a "corkscrew" sequence of turning point bifurcations while their period (not shown in the figure) approaches infinity. In the figure, the branch is seen to asymptotically approach the bi–tri steady state solution branch. This behavior is typical of a homoclinic connection to a saddle–focus (a Silnikov loop). A detailed discussion of the bifurcation behavior associated with this phenomenon in three space dimensions (a one dimensional unstable manifold connecting with a two dimensional stable manifold corresponding to a complex eigenvalue pair) can be found in [18] (see also [17]). A space–time plot and a Fourier space projection of a representative limit cycle close to the "tip" of the corkscrew (figures 7b and c) illustrate the geometry of this connection: the limit cycle is indeed seen to approach (and spend most of its period in the neighborhood of) the bi–tri saddle type steady state. The nature of the eigenvalues of this steady state (confirmed by independent computations) is also obvious in the figure: one is positive and real, while the two least stable eigenvalues of the saddle form a complex conjugate pair (indicated by the spiraling close to the steady state). The limit cycle branch is expected to corkscrew on its approach to homoclinicity if the Silnikov condition, $|\delta| < 1$, is satisfied; δ is the ratio of the real part of the complex eigenvalue pair to the real eigenvalue of the saddle–focus (see [18]). A number of other features, qualitatively and quantitatively predicted in the Glendinning and Sparrow analysis have also been computationally observed, and are partly seen in this figure (for a better illustration see figure 10). These include the (asymptotically exponential) narrowing of the parameter interval between successive turning points on the corkscrew, the existence of period doubling bifurcations and their relative location with respect to the turning points as well as subsidiary homoclinicities.

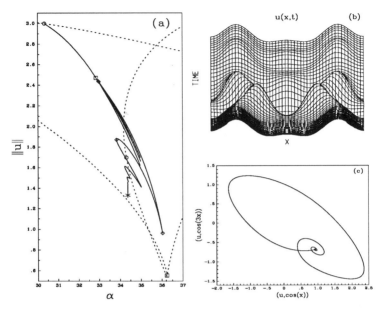

Figure 7: (a) Partial bifurcation diagram of the asymmetric limit cycle branch. (b) Space–time plot and (c), an a_1–a_3 phase space projection of a limit cycle approaching homoclinicity ($\alpha \sim 34.37$, marked with a *).

Figure 8a overlays the fate of the initial period–doubled and the subsequent period–quadrupled limit cycle branches on figure 7a. We observe that they also asymptotically approach corkscrews located close to that of the primary asymmetric oscillations. Figures 8b and c show Fourier space projections of representative limit cycles close to the tips of the corresponding corkscrews, indicating again an interaction between the stable and unstable manifolds of the same branch of bi–tri saddle type steady state solutions. Period doubling bifurcations like those on the primary branch are also expected close to the limit points of these subsidiary corkscrews.

Figure 9 shows the fate of the original symmetric limit cycle branch. Computations of this branch also display the beginnings of a corkscrew behavior. However, as the Fourier space projection portraits of representative limit cycles on this branch indicate (figures 9b and c), the limit cycles appear to asymptote towards a symmetric heteroclinic connection. This connection apparently involves the unstable manifold of one saddle type steady state with the stable manifold of its π shift and vice-versa. The computation of this branch presents certain numerical difficulties since the computed limit cycles are very unstable (typical Floquet multipliers even before the first turning point become so large $-O(10^{10})$ and more– as to be inaccurate). These phase portraits are strongly reminiscent of the structure of the Lorenz attractor, and the bifurcation behavior of orbits of this nature has been discussed in Sparrow's book [37].

A much more "textbook" example of Silnikov type behavior is exhibited by the limit cycle branches emanating from the Hopf bifurcation point at $\alpha \sim 34.30$ on the bi–tri steady state branch. The two representatives of this branch have the same stability in the C_1 space and therefore two Hopf bifurcations occur simultaneously,

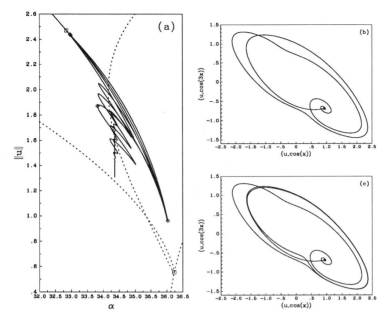

Figure 8: (a) Partial bifurcation diagram of the asymmetric limit cycle branches. a_1–a_3 phase space projections of the period–2 (b) and period–4 (c) asymmetric limit cycles close to homoclinicity ($\alpha \sim 34.36$).

giving rise to two (symmetric to each other) limit cycle branches. Because there are no symmetry restrictions, these branches can (and indeed are observed to) period double according to the typical Silnikov scenario. An initial period doubling at $\alpha \sim 34.54$ results in a limit cycle branch which, as seen in figure 10a, apparently asymptotically approaches a Silnikov connection of a saddle focus on the "middle" (in the figure) steady state branch. A second period doubling at $\alpha \sim 35.52$ gives rise to a branch with a similar fate. Interestingly the next period doubling on the "primary" corkscrew ($\alpha \sim 34.84$) gives rise to a period doubled branch which appears to asymptotically approach a homoclinic connection of a saddle lying on a different part of the steady state branch. This is a "subsidiary homoclinicity" and a representative limit cycle portrait on this branch (figure 10c) clearly illustrates closeness to a "double pass" Silnikov connection (as opposed to the "single pass" of the primary homoclinicity (figure 10b).

Obviously the details of these diagrams are incomplete –one could go on computing limit points, period doublings, and subsequent homoclinicities forever. It is sufficient to summarize this entire scenario as a form of "loss of a limit cycle branch" [18]. Before concluding this descriptive section, we should add that even this complicated scenario does not contain all essential features of the dynamics of the KSE in C_1 in the parameter interval studied; Simulations indicate additional global bifurcations for even earlier values of α than the ones discussed here (e.g. close to $\alpha = 32.9000355$); the picture is far from being complete.

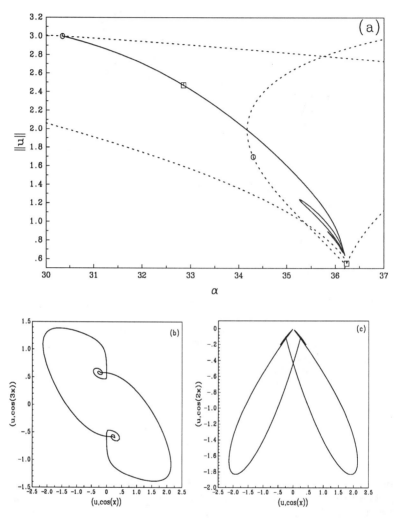

Figure 9: (a) Partial bifurcation diagram of the symmetric limit cycle branch. (b) a_1–a_3 and (c) a_1–a_2 phase space projections of a representative limit cycle ($\alpha \sim 36.15$).

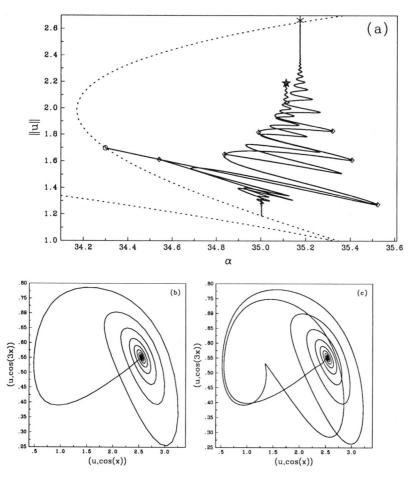

Figure 10: (a) Partial bifurcation diagram of the limit cycle branches bifurcating from the bi–tri steady state branch. (b) and (c) show phase space projections of a period 1 ($\alpha \sim 35.17$, marked with a *) and a period 2 ($\alpha \sim 35.11$, marked with a $\star$) limit cycle approaching homoclinicity.

3. An accurate three mode model. The bifurcation calculations in the previous section were performed using a nine mode traditional Galerkin spatial discretization of the PDE. While the computational task involved in this level of analysis for nine ODEs is tractable, it is conceivable that an alternative discretization technique could capture the dynamics accurately with a smaller number of modes. A traditional approach to model reduction, particularly suitable for the qualitative local analysis of bifurcating branches, involves studying the flow on a center manifold in the neighborhood of a particular bifurcating solution. An example of this type of analysis for the KSE in the recent literature can be found in [1]. An alternative approach involves Galerkin projection on empirically determined basis functions obtained from statistical processing of computationally or experimentally obtained time series. This more practically oriented approach (method of empirical eigenfunctions or Proper Orthogonal Decomposition, POD), introduced by Lumley [30, 31] as a method of detecting spatially coherent structures in turbulence, has been extensively used in recent years for the study of transitional flows in various geometries (e.g. [4, 35, 11]) and references therein).

A *global* (in phase space) model reduction approach is motivated by the theory of inertial manifolds. This theory can be described for a dissipative PDE written as an evolutionary equation in a Hilbert space H

$$\frac{du}{dt} + Au + F(u) = 0, \quad u \in H.$$

For example, for the KSE we let $A = \frac{\partial^4}{\partial x^4}$ and

$$F(u) = \alpha \left(\frac{\partial^2 u}{\partial x^2} + \frac{1}{2} \left(\frac{\partial u}{\partial x} \right)^2 - \frac{1}{4\pi} \int_0^{2\pi} (u_x)^2 \, dx \right).$$

An inertial manifold is a finite dimensional, positively invariant, exponentially attracting Lipschitz manifold embedded in the phase space H. Let $P = P_m$ denote the orthogonal projector onto the span of the first m eigenfunctions of A, and let $Q = Q_m = I - P_m$ be that onto the orthogonal complement. Typically an inertial manifold is constructed as the graph of a function $\Phi : PH \to QH$. This function is used to define a finite dimensional ODE (the inertial form)

$$\frac{dp}{dt} + Ap + PF(p + \Phi(p)) = 0, \quad p \in PH$$

which captures all the long time behavior of the PDE.

While inertial manifolds are known to exist for the KSE as well as other dissipative PDEs (see e.g. Constantin *et al.* [10]), they cannot in general be expressed in closed form. For this reason a number of approximate inertial manifolds (AIMs) have been introduced in the literature (see [32] and the references therein). The AIM used here is designed to pass near all steady states of the PDE, which lie on the global attractor, and consequently on any inertial manifold. This construction is also well motivated for dissipative PDEs that may not necessarily possess an inertial manifold. Indeed, this AIM was originally introduced in [39] for the Navier-Stokes equations (NSE). Consider the Q-component of the condition for a steady state

$$Aq + QF(p + q) = 0, \quad p \in PH, \ q \in QH,$$

which implicitly defines an analytic function $\Phi^* : PH \to QH$ given by the fixed points for the family of mappings

$$T_p(q) = -A^{-1}QF(p+q),$$

provided m is large enough. The AIM used in computations in this paper

$$\Phi_2(p) \equiv T_p(T_p(0))$$

has a distance to Φ^* (in L^2) which is comparable to that between Φ^* and the global attractor (see [39] for the NSE, [23] for the KSE).

The implementation of this nonlinear Galerkin approximation is well suited for computation using spectral methods. The nonlinear Galerkin method used here i.e. the m dimensional ODE system

$$(3) \qquad \frac{dp}{dt} + Ap + PF(p + \Phi_2(p)) = 0, \qquad p \in PH,$$

(called an Approximate Inertial Form, AIF) amounts to three successive evaluations of the nonlinearity F. For the KSE, with its quadratic nonlinearity and periodic boundary conditions, the graph of Φ_2 lives in the span of the first $4m$ eigenfunctions of A (Fourier modes, cosines for the even functions here). This means that with sufficient padding with zeros, spectral codes can be used to compute the vectorfield of (3) faithfully. In the general case where the nonlinearity is not of polynomial type, a collocation scheme can be implemented efficiently using the Fast Fourier Transform, albeit with aliasing errors [7].

Obviously, the minimal number of modes (phase space dimension) necessary to accommodate the Silnikov loop phenomena described above, is three. However, a traditional three mode Galerkin discretization does not capture these phenomena [23]. The best *a priori* estimates for the dimension of an inertial manifold for the KSE do not suggest that this dimension could be as low as three or, for that matter, even nine (which we empirically know to be sufficient from our computational results). Our choice here of the dimension of the approximate inertial manifold (three) is therefore based on the dynamic phenomena observed rather than on rigorous estimates.

Figure 11 shows the relevant section of the bifurcation diagram for the AIF (3) with three modes, which does indeed exhibit the same limit cycle bifurcation behavior as the "converged" diagram (figure 6). The original Hopf bifurcation from the bimodal branch to a symmetric limit cycle, the subsequent symmetry breaking followed by period doublings, as well as the ultimate Silnikov loop terminations of the resulting branches are present (we had some numerical difficulties with the extensive continuation of the symmetric limit cycle branch). Similarly, the "nonsymmetric" limit cycle branches bifurcating off of the bi–tri mixed mode steady state branch, their subsequent period doublings, Silnikov loops, and subsidiary homoclinicities are correctly (and, to some extent, accurately) reproduced. The actual parameter values at which the Silnikov connections occur in the reduced system are not identical to the traditional Galerkin values; however, they are in reasonable semi–quantitative agreement.

4. Approximating the Silnikov connections.
The existence of certain Silnikov type homoclinic connections "implies" the existence of a complicated pattern

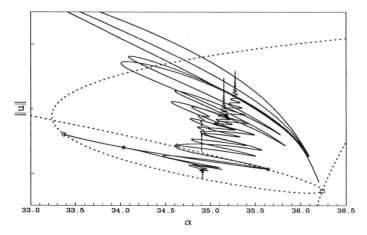

Figure 11: Partial bifurcation diagram for the KSE obtained using the AIF (3) with $m = 3$.

of bifurcations and oscillatory branches in the neighborhood of these connections in (phase × parameter) space. Several aspects of this scenario were illustrated above for the KSE. On the other hand, this entire bifurcation sequence can be thought of as a mechanism for the disappearance of the "primary" limit cycle branch as the bifurcation parameter changes. For example, the collection of limit cycle branches around $\alpha \sim 35.17$ in figure 10 can be taken to simply denote the "death" of the nonsymmetric limit cycle branch originating at $\alpha \sim 34.30$.

The continuation of limit cycle branches as they approach homoclinicity involves considerable computational effort without necessarily producing significant information. Due to its infinite complexity, the diagram can never be completed; in addition, as the limit cycle branches approach homoclinicity their period approaches infinity, and their accurate calculation poses severe discretization problems. On the other hand, all these complications occur in a very narrow parameter interval, and it appears that approximating the Silnikov connection directly would for all practical purposes contain the essential information regarding the behavior of the limit cycle branch(es).

Figure 7c is representative of the phase space geometry of a Silnikov connection. The one dimensional unstable manifold of a saddle steady state connects with the saddle itself (and therefore lies on the stable manifold). The spiraling approach to the saddle is indicative of the complex conjugate nature of the two least stable saddle eigenvalues ("saddle–focus"). This figure is of course only a projection, and the stable manifold is not two dimensional, as was the case in Glendinning and Sparrow (for this figure it is actually eight dimensional). The model reduction approach discussed in section 3 reduces the dimensionality of the stable manifold to two, the minimum necessary, and allows for the easy visualization of both invariant manifolds and their interactions. We have used this reduction to implement a geometrically motivated simple shooting algorithm to converge on the Silnikov loop. Such algorithms (special cases of boundary value problems on semi–infinite or infinite intervals [29]) have appeared in the literature for the calculation of saddle connections in low dimensional

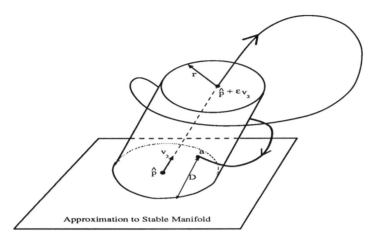

Figure 12: Illustration of the geometry and terminology used in the shooting algorithm.

systems [20] (see [28] for the Silnikov case). More recently, results for the general case with arbitrary (finite) dimensional interacting manifolds have been reported [14, 15, 16, 6, 5].

We now present an outline of our implementation for the case when the associated steady state has a one dimensional unstable manifold and a two dimensional stable manifold. For simplicity we rewrite the approximate inertial form (3) as

(4) $$\dot{p} = G(p; \alpha).$$

Let $\hat{p}(\alpha)$ denote a fixed point of (4) with associated stable eigenvector $v_1 \pm iv_2$ and unstable eigenvector v_3. Let C_r define a cylinder of radius r with its axis in the direction of v_3 passing through the point $\hat{p}$. Consider also the plane spanned by the vectors v_1, v_2 passing through $\hat{p}$ (see figure 12). Given a good guess α_n for the parameter value, α^*, at which the Silnikov loop occurs, the computational procedure is outlined below:

1. Locate the steady state $\hat{p}(\alpha_n)$ via Newton iteration and compute its eigenvectors.

2. Integrate

$$\dot{p} = G(p; \alpha_n)$$
$$p(t = 0) = \hat{p} + \epsilon v_3$$

   until the trajectory intersects the cylinder C_r as shown in figure 12. Let a denote the intersection point.

3. Calculate the distance $D = D(\alpha_n, \epsilon, r)$ from a in the direction of v_3 to the plane. To satisfy $D(\alpha) = 0$:

4. Calculate the derivative $\frac{dD}{d\alpha}$ and use it to update the current guess for α via a Newton iteration:

$$\alpha_{n+1} = \alpha_n - \left[\frac{dD}{d\alpha}\right]^{-1}_{\alpha_n} D|_{\alpha_n}.$$

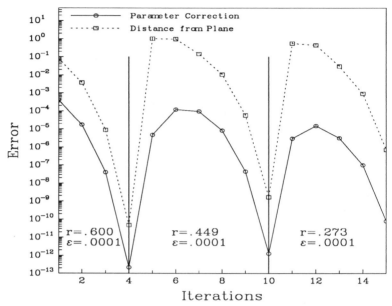

Figure 13: Rate of convergence to the Silnikov connection at $\alpha \sim 35.2741195$.

Upon convergence of this procedure we obtain an $\alpha(\epsilon, r)$ satisfying $D(\alpha, \epsilon, r) \sim 0$. Since v_3 is tangent to the unstable manifold at $\hat{p}$, it is obviously a good approximation of the unstable manifold for small enough ϵ. Similarly, the plane provides a good approximation of the stable manifold for small r. Therefore, in the limit, $\alpha(r, \epsilon)$ should approximate α^*. In our implementation we have repeated the above procedure by systematically shrinking r and ϵ until $\alpha(\epsilon, r)$ remains constant to within a prescribed error tolerance (usually $< 10^{-8}$).

Calculation of the intersection point a involves a Newton–Raphson iteration ensuring that it lies on C_r, which implicitly defines the "time of flight" of the trajectory, and can be used to calculate its dependence on the system parameters. The most involved segment of the computational procedure is the evaluation of the derivative $\frac{dD}{d\alpha}$. D depends on α both explicitly through the vectorfield and implicitly through the initial condition (the fixed point and its eigenvectors) and the cylinder itself (again depending on the eigenvectors). The calculation therefore involves repeated application of the chain rule, using information from integration of the sensitivity equations (dependence of the flow on α). It is of course possible to use numerical derivatives in this calculation. Because of the complexity of the implicit differentiations involved in computing the "true" derivatives, we preferred to use *linear* approximations to the stable and unstable manifolds, and progressively "squeeze" them very close to the saddle (where they become increasingly accurate), as opposed to higher order approximations to the local manifolds as in [20, 28]; these approximations are more accurate further away from the saddle, but evaluating their dependence on the parameter becomes increasingly complicated.

Figure 13 shows the convergence rate of the Newton iteration for the location of the Silnikov connection at $\alpha \sim 35.2741195$. The "elbows" in the graph correspond to changes in r. The figure has been obtained by letting $D(\alpha, \epsilon, r)$ approach zero to within 10^{-6} before shrinking the cylinder. Upon convergence of the iteration for a prescribed value of r, the new r has been chosen by allowing the last computed

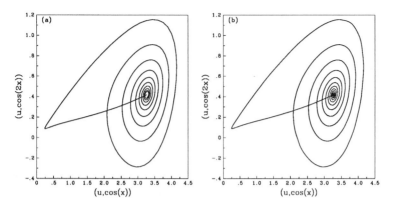

Figure 14: Phase space projections of a converged Silnikov connection ($\alpha = 35.2741195$) and a limit cycle near the the tip of the corkscrew ($\alpha = 35.27398$).

trajectory to further approach the fixed point as close as possible before it starts moving away from it. The reason for not immediately starting with a very small (accurate) r is that simple shooting may fail to return to the cylinder if the initial guess of α is not very accurate. This probably means that convergence for multiple shooting or collocation discretizations for the boundary value problem would be easier. Friedman and Doedel, in a series of papers [14, 15, 16], have discussed various issues associated with computing homoclinic and heteroclinic connections between steady states in a more general setting. Their implementation of these algorithms in a version of AUTO will undoubtedly make such calculations routinely possible. Figure 14a shows our approximation of the single pass Silnikov connection ($\alpha \sim 35.2741195$) and 14b shows a nearby limit cycle computed by AUTO (in this case the unstable manifold was actually two dimensional and the stable manifold was one dimensional, but the Silnikov condition was still obeyed).

Summary and conclusions. A detailed discussion of the spatio–temporal patterns associated with certain limit cycle solution branches of the KSE has been presented for the parameter interval $0 < \alpha < 40$. A particular space–time symmetry ("Ponies On a Merry-go-round") of one of these branches, its origin and its effect on subsequent bifurcations has also been discussed. We computationally observe that all limit cycle branches studied apparently "terminate" in homoclinic (or heteroclinic) Silnikov connections, involving saddle type steady state solutions. Numerical evidence for the occurrence of a number of features generically associated with a Silnikov loop (period doublings, subsidiary homoclinicities, etc.) has also been presented. The minimal phase space dimension consistent with these observations is three; yet a traditional three mode Galerkin truncation of the PDE does not capture this behavior. Exploiting the theory of Approximate Inertial Manifolds (AIMs), we were able to construct a minimal (three dimensional) model capable of qualitatively (and almost quantitatively) reproducing the bifurcation structure.

The sharp reduction in dimensionality allows for the easier visualization of the (approximate) long term dynamics of the PDE; in our three dimensional model the highest dimension of the interacting stable and unstable manifolds of the saddle–focus is two. We exploited this tractable geometry in implementing a simple shooting algorithm to accurately locate the Silnikov connections. More generally, accurate model reduction for PDE discretizations is important in allowing the approximation of (low–dimensional) stable manifolds of steady or oscillatory solutions further away from these solutions in phase space. This should allow for an easier geometric under-standing of the phase space structures underlying long term dynamics of the PDE, especially global bifurcations. Our particular example, the KSE, is very well suited for model reduction using AIMs: dissipation is strong, the first few eigenfunctions of the linear "dissipative" part, used to parameterize the AIM, also happen to be the eigendirections in which the first few instabilities occur and carry most of the energy of the long term solutions. In addition, the spectral implementation of the AIM ben-efits from the fact that these eigenfunctions are pure Fourier modes. For other PDEs and other geometries, it is quite possible that alternative basis functions (e.g. based on some empirical information about the problem, as is the case in POD) can be much more efficient in parametrizing an appropriate low–dimensional manifold, and hence give rise to an accurate dynamical system of smaller dimension.

Acknowledgements: We are deeply indebted to Professor Edriss S. Titi for his invaluable contributions to many aspects of this research. The work of H.S.B. and I.G.K. was supported in part by NSF Grants CTS–8957213, DMS–8906292 and a David and Lucile Packard Foundation Fellowship; that of M.S.J. was supported in part by NSF grant DMS–9007802. Part of this work was performed while I.G.K. was enjoying the hospitality (and M.S.J. was a postdoctoral member) of the Institute for Mathematics and its Applications at the University of Minnesota.

References

[1] D. Armbruster, J. Guckenheimer, and P. Holmes. Kuramoto–Sivashinsky dynamics on the center–unstable manifold. *SIAM J. Appl. Math.*, 49(3):676–691, June 1989.

[2] D. G. Aronson, M. Golubitsky, and J. Mallet-Paret. Ponies on a merry-go-round in large arrays of Josephson junctions. *Nonlinearity (to appear)*, 1991.

[3] P. J. Aston. Introduction to the numerical solution of symmetry–breaking bifur-cation problems. In D. Roose, B. DeDier, and A. Spence, editors, *Proceedings on the NATO Advanced Research Workshop on Continuation and Bifurcations: Numerical Techniques and Applications*, pages 139–152. Kluwer Academic Pub-lishers, 1990.

[4] N. Aubry, P. Holmes, J. Lumley, and E. Stone. The dynamics of coherent struc-tures in the wall region of a turbulent boundary layer. *J. F. M.*, 192:115–173, 1988.

[5] W. J. Beyn. Global bifurcations and their numerical computation. In D. Roose, B. DeDier, and A. Spence, editors, *Proceedings on the NATO Advanced Research Workshop on Continuation and Bifurcations: Numerical Techniques and Appli-cations*, pages 169–181. Kluwer Academic Publishers, 1990.

[6] W. J. Beyn. The numerical computation of connecting orbits in dynamical systems. *IMA J. Num. Anal.*, 9:379–405, 1990.

[7] H. S. Brown, M. S. Jolly, I. G. Kevrekidis, and E. S. Titi. Use of approximate inertial manifolds in bifurcation calculations. In D. Roose, B. DeDier, and A. Spence, editors, *Proceedings on the NATO Advanced Research Workshop on Continuation and Bifurcations: Numerical Techniques and Applications*, pages 9–23. Kluwer Academic Publishers, 1990.

[8] H. S. Brown and I. G. Kevrekidis. A numerical study of homoclinic orbits to a saddle–focus and dynamics associated with their existence. *Paper 187e, Presented to the 1990 AIChE annual meeting, Chicago*, 1990.

[9] L. H. Chen and H. C. Chang. Nonlinear waves on thin film surfaces. II. bifurcation analysis of the long–wave equation. *Chem. Eng. Sci.*, 41:2477–2486, 1986.

[10] P. Constantin, C. Foias, B. Nicolaenko, and R. Témam. *Integral Manifolds and Inertial Manifolds for Dissipative Partial Differential Equations.* Appl. Math. Sci., No. 70. Springer–Verlag, 1988.

[11] A. E. Deane, I. G. Kevrekidis, G. E. Karniadakis, and S. A. Orszag. Low dimensional models for complex geometry flows: Application to grooved channels and circular cylinders. *Phys. Fluids A (Submitted)*, 1990.

[12] E. S. Doedel. AUTO: A program for the automatic bifurcation analysis of autonomous systems. *Cong. Num.*, 30:265–284, 1981.

[13] E. S. Doedel. *AUTO 86 User Manual*, February 1986.

[14] E. S. Doedel and M. J. Friedman. Numerical computation of heteroclinic orbits. *J. Comput. and Appl. Math.*, 26:159–170, 1989.

[15] M. J. Friedman and E. Doedel. Numerical computation and continuation of invariant manifolds connecting fixed points with application to computation to combustion fronts. In *Proc. 7th Int. Conf. on Finite Element Methods in flow problems*, April 1989.

[16] M. J. Friedman and E. Doedel. Numerical computation and continuation of invariant manifolds connecting fixed points. *SIAM J. Num. Anal.*, 1991. to appear.

[17] P. Gaspard, R. Kapral, and G. Nicolis. Bifurcation phenomena near homoclinic systems: A two–parameter analysis. *J. Stat. Phys.*, 35(5/6):697–727, 1984.

[18] P. Glendinning and C. Sparrow. Local and global behavior near homoclinic orbits. *J. Stat. Phys.*, 35:645–696, 1984.

[19] J. M. Greene and J. S. Kim. The steady states of the Kuramoto–Sivashinsky equation. *Physica D*, 33:99–120, 1988.

[20] B. Hassard. Computation of invariant manifolds. In P. Holmes, editor, *New Approaches to Nonlinear Problems in Dynamics*, pages 27–42. SIAM, 1980.

[21] J. M. Hyman, B. Nicolaenko, and S. Zaleski. Order and complexity in the Kuramoto–Sivashinsky model of weakly turbulent interfaces. *Physica D*, 23:265–292, 1986.

[22] J. S. Il'yashenko. Global analysis of the phase portrait for the Kuramoto–Sivashinsky equation. Technical Report 665, IMA Preprint, 1990.

[23] M. S. Jolly, I. G. Kevrekidis, and E. S. Titi. Approximate inertial manifolds for the Kuramoto–Sivashinsky equation: Analysis and computations. *Physica D*, 44:38–60, 1990.

[24] I. G. Kevrekidis, H. S. Brown, and M. S. Jolly. Low–dimensional chaos in models of interfacial instabilities: Global bifurcations and the role of unstable solutions. *Paper 16e, Presented to the 1989 AIChE annual meeting, San Francisco*, 1989.

[25] I. G. Kevrekidis and M. S. Jolly. On the computation of inertial manifolds. *Paper 173c, Presented to the 1988 AIChE annual meeting, Washington*, 1988.

[26] I. G. Kevrekidis, B. Nicolaenko, and J. C. Scovel. Back in the saddle again: A computer assisted study of the Kuramoto–Sivashinsky equation. *SIAM J. Appl. Math.*, 50(3):760–790, 1990.

[27] Y. Kuramoto and T. Tsuzuki. Persistent propagation of concentration waves in dissipative media far from thermal equilibrium. *Prog. Theor. Phys.*, 55(2):356–369, February 1976.

[28] Y. A. Kuznetsov. Computation of invariant manifold bifurcations. In D. Roose, B. DeDier, and A. Spence, editors, *Proceedings on the NATO Advanced Research Workshop on Continuation and Bifurcations: Numerical Techniques and Applications*, pages 183–195. Kluwer Academic Publishers, 1990.

[29] M. Lentini and H. B. Keller. Boundary value problems on semi–infinite intervals and their numerical solution. *SIAM J. Num. Anal.*, 17(4):577–604, August 1980.

[30] J. L. Lumley. The structure of inhomogeneous turbulent flows. In A. M. Yaglom and V. I. Tatarski, editors, *Atmospheric Turbulence and Radio Wave Propagation*, pages 166–178. Moscow: Nauka, 1967.

[31] J. L. Lumley. *Stochastic Tools in Turbulence*. Academic Press, NY, 1970.

[32] M. Luskin and G. Sell. Approximation theories for inertial manifolds. Technical Report 88/76, University of Minnesota Supercomputer Institute, July 1988.

[33] B. Nicolaenko, B. Scheurer, and R. Témam. Some global dynamical properties of the Kuramoto Sivashinsky equation: Nonlinear stability and attractors. *Physica D*, 16:155–183, 1985.

[34] J. C. Scovel, I. G. Kevrekidis, and B. Nicolaenko. Scaling laws and the prediction of bifurcations in systems modeling pattern formation. *Phys. Letters A*, 130:73–80, 1988.

[35] L. Sirovich and C. Sirovich. Low dimensional description of complicated phenomena. *Contemporary Mathematics*, 89:277–365, 1989.

[36] G. I. Sivashinsky. Nonlinear analysis of hydrodynamic instability in laminar flames–I. derivation of basic equations. *Acta Astronautica*, 4:1177–1206, 1977.

[37] C. Sparrow. *The Lorenz Equations: Bifurcations, Chaos and Strange Attractors.* Appl. Math. Sci., No. 41. Springer–Verlag, 1982.

[38] J. W. Swift and K. Wiesenfeld. Suppression of period doubling in symmetric systems. *Phys. Rev. Letters*, 52(9):705–708, February 1984.

[39] E. S. Titi. On approximate inertial manifolds to the Navier–Stokes equations. *J. Math. Anal. Appl.*, 149:540–557, 1990.

LOCALIZED AND EXTENDED PATTERNS IN REACTIVE MEDIA

CHRISTIAN ELPHICK* AND EHUD MERON†

Abstract. We study the dynamics of interacting localized structures in homogeneous reactive media. Equations of motion for solitary waves in excitable media and for vortices in oscillatory media are derived under the assumption of weak interactions. We show that excitable media with oscillatory recovery can support a multitude of stable, nonuniform spatial patterns and that phase field effects in oscillatory media may lead to the formation of bound vortex pairs. The implications of the latter result on the transition to turbulence in oscillatory media are discussed.

1. Introduction. Reactive media provide important case studies for pattern formation problems in far from equilibrium systems. Reactive solutions with excitable chemical kinetics can be used as homogeneous and isotropic model systems in the study of various wave phenomena in biological systems [1,2], whereas oscillating chemical reactions are useful for studying phase dynamics and weak turbulence [3].

A significant feature of reactive media is the existence of localized spatial structures. The simplest of these are the one-dimensional solitary waves of chemical activity that propagate in quiescent excitable media. These waves constitute the chemical analog of action potentials in biological membranes [1,2,4]. In two space dimensions vortex-like structures are known to exist which take the form of rotating spiral waves [5]. Very often, extended patterns can be viewed as systems of many interacting localized ones. We will adopt here this approach but consider only sparse patterns in which collisions among localized structures are avoided.

In addressing the problem of interacting localized structures two questions arise: what degrees of freedom are relevant for the asymptotic time evolution, and how to derive their dynamics in the presence of perturbations. We will see in section 2 that the key to these questions rests in the continuous symmetries of the system. The potential degrees of freedom of a localized pattern that breaks these symmetries are the corresponding symmetries parameters (position in the case of translational symmetry, etc.). In the presence of symmetry breaking perturbations these parameters become slow dynamical variables. Evolution equations for these variables appear as solvability conditions which guarantee that the solution of the unperturbed problem, but with time dependent symmetry parameters, is an approximate solution of the perturbed equations. In section 3 we will use these ideas to derive the time evolution of wavetrains of impulses that propagate along one-dimensional media, and show how spatially complex patterns may emerge in excitable systems with oscillatory recovery.

There are situations in which the interactions among localized structures are mediated by extended fields. An example will be presented in section 4 where the

*Physics Department, Universidad Tecnica F. Santa Maria, Casilla 110-V, Valparaiso, Chile.
†Department of Chemical Physics, Weizmann Institute of Science, Rehovot 76100, Israel.

dynamics of vortices in oscillating, two-dimensional media will be studied. It will be shown in this section that phase field effects may lead to the formation of bound vortex pairs. Possible implications of this result on the transition to turbulence will be discussed in section 5.

2. Dynamics of Localized Structures. Reactive systems, in the absence of convection and thermal effects, are normally described by reaction-diffusion equations (rde's) of the form

$$(2.1) \qquad \partial_t \mathbf{U} = \mathbf{R}(\mathbf{U}) + \mathbf{D}\nabla^2 \mathbf{U},$$

where $\mathbf{U}$ is a set of concentration fields, $\mathbf{R}(\mathbf{U})$ represents the mass-action kinetics and $\mathbf{D}$ is a matrix of diffusion constants. For the sake of illustration we confine ourselves in this section to patterns in one space dimension. We assume that (2.1) has an equilibrium solution, $\mathbf{U} = \mathbf{0}$, and solutions which describe localized patterns such as solitary waves that propagate at constant speed, $\mathbf{U}(x,t) = \mathbf{H}(x - c_0 t)$. We also assume that (1.1) describes a homogenous medium and therefore invariant under translations. The latter assumption implies that the translated form $\mathbf{H}(\chi - \chi_0)$, where $\chi := x - c_0 t$, is also a solution of (2.1).

Consider now a perturbation, $\mathbf{P}$, that breaks the translational invariance of the system. The perturbation can be external or due to other localized structures in the neighborhood. We will consider in this section the former case, postponing the many body problem of interacting structures to the next section. The main effect of such a perturbation is to make the symmetry parameter χ_0 a slow dynamical variable, $\chi_0(\epsilon t)$, where ϵ is a small parameter that measures the size of the perturbation. The rational for that derives from the marginal stability of $\mathbf{H}$ to translational perturbations. An equation of motion for $\chi_0(\epsilon t)$ now follows from the requirement that $\mathbf{H}(\chi - \chi_0(\epsilon t))$ is an approximate solution of the perturbed problem

$$(2.2) \qquad \partial_t \mathbf{U} = \mathbf{R}(\mathbf{U}) + \mathbf{D}\partial_x^2 \mathbf{U} + \epsilon \mathbf{P}.$$

To see this we write the solution in the form

$$(2.3) \qquad \mathbf{U} = \mathbf{H}(\chi - \chi_0(\epsilon t)) + \epsilon \mathbf{C}(\chi, \epsilon t),$$

insert it in (2.2) and evaluate an equation for the correction term $\mathbf{C}$. To leading order in ϵ we obtain

$$(2.4) \qquad \mathcal{L}\mathbf{C} = \chi_0' \mathbf{H}' + \mathbf{P} \qquad \mathcal{L} := -c_0 \partial_\chi - \mathbf{D}\partial_x^2 - \nabla_\mathbf{H}\mathbf{R}(\mathbf{H}),$$

where the prime denotes differentiation with respect to the argument. In order for $\mathbf{H}(\chi - \chi_0(\epsilon t))$ to be an approximate solution of (2.2), solutions $\mathbf{C}$ of (2.4) should be finite. We are thus led to the search of zero eigenmodes (or null vectors) of $\mathcal{L}$. If such eigenmodes exist, solvability conditions should be employed to eliminate singular components from $\mathbf{C}$. It is easy to see that $\mathcal{L}$ has indeed a zero eigenmode. The solitary-wave solution $\mathbf{H}(\chi)$ satisfies the unperturbed equation (2.1). Differentiating this equation with respect to $\chi = x - c_0 t$ we find

$$(2.5) \qquad\qquad \mathcal{L}\mathbf{H}' = 0.$$

We may call $\mathbf{H}'$ the translation mode since translational perturbations, $\mathbf{H}(\chi + \delta\chi)$, are along that mode. Equation (2.5) suggests the existence of a zero eigenmode of the adjoint operator as well [6],

$$(2.6) \qquad\qquad \mathcal{L}^{\dagger}\mathbf{G} = 0,$$

where the adjoint property is defined with respect to an appropriately chosen inner product. Taking the inner product of (2.6) with the correction term $\mathbf{C}$ we find

$$(2.7) \qquad\qquad \dot{\chi}_0 = -(\mathbf{G}, \epsilon\mathbf{P})/(\mathbf{G}, \mathbf{H}'),$$

where the overdot denotes differentiation with respect to time. An equation of motion for χ_0 follows once we realize that $\mathbf{G}(\chi - \chi_0)$ is localized around $\chi = \chi_0$ and $\mathbf{P}$ is an explicit function of x. This allows us to express the right hand side (rhs) of (2.7) as a function of χ_0, a function which reflects the form of the symmetry breaking perturbation $\mathbf{P}$.

We have considered a system that has only one continuous symmetry, the symmetry of translations. As a result the asymptotic dynamics of a localized pattern in such a system is governed by a first order, ordinary differential equation (ode) for the corresponding symmetry parameter χ_0. The method introduced here can be extended to systems possessing a number of continuous symmetries. The time evolution of structures that break these symmetries will be described by sets of first order ode's for the pertinent symmetry parameters. An example of this richer case has recently been studied [7].

3. Extended Patterns in One Space Dimension. In the previous section we considered the dynamics of a localized structure in the presence of an external perturbation $\mathbf{P}$. We proposed an approximate solution of the perturbed problem (2.2) in the form (2.3) where the time evolution of $\chi_0(\epsilon t)$ is determined by (2.7). In this section we consider the dynamics of widely spaced wavetrains of solitary waves or impulses. Here, even in the absence of external perturbations, one may expect for non-trivial dynamics, for impulses are perturbed by the fields of their neighbors.

To describe the dynamics of widely spaced wavetrains we look for solutions of (2.1) of the form

$$(3.1) \qquad\qquad \mathbf{U} = \sum_j \mathbf{H}\big(\chi - \chi_j(\epsilon t)\big) + \epsilon\mathbf{C}(\chi, \epsilon t),$$

where ϵ is a measure of the perturbations that are exerted by nearby impulses. To be more specific about the small parameter ϵ we need to consider the asymptotic forms of the solitary wave solution, $\mathbf{H}(\chi)$, as $\chi \to \pm\infty$. These forms are generically exponential functions for they follow from a linear analysis of (2.1) around the equilibrium state $\mathbf{U} = \mathbf{0}$. We thus write

$$(3.2a) \qquad \mathbf{H}(\chi) \sim e^{-\eta_R \chi} \cos(\nu_R \chi + \phi_R) \; \chi \to \infty,$$

$$(3.2b) \qquad \mathbf{H}(\chi) \sim e^{\eta_L \chi} \cos(\nu_L \chi + \phi_L) \; \chi \to -\infty.$$

We can define now our small parameter to be $\epsilon = \exp(-\eta\lambda)$ where $\eta = \min(\eta_R, \eta_L)$ and λ is a typical spacing between successive impulses. Clearly, in the limit of infinite spacings, $\epsilon = 0$ and the form (3.1) reduces to the solitary solution $\mathbf{H}$.

Equations of motion for the impulse positions χ_j (in a frame moving with velocity c_0) follow from solvability conditions which guarantee the finiteness of the correction term $\mathbf{C}$. The derivation goes along the lines depicted in the previous section and the result takes a form similar to that of (2.7):

$$(3.3) \qquad \dot{\chi}_j = -(\mathbf{G}_j, \epsilon\mathbf{P})/(\mathbf{G}_j, \mathbf{H}'_j),$$

where $\mathbf{P}$ is given by

$$(3.4) \qquad \epsilon\mathbf{P} = \mathbf{R}(\sum_j \mathbf{H}_j) - \sum_j \mathbf{R}(\mathbf{H}_j),$$

and $\mathbf{F}_j := \mathbf{F}(\chi - \chi_j)$. In (2.7) $\mathbf{P}$ stands for a symmetry breaking *external* perturbation. Here, it represents spatial nonhomogeneities generated by the impulses themselves. Notice that only nonlinear terms in $\mathbf{R}(\mathbf{U})$ contribute to (3.4) and that at any point in space the rhs of (3.4) is at most of $O(\epsilon)$. To evaluate the rhs of (3.4) we need to define an inner product. We assume the usual inner product of vector analysis and use integration over the infinite range of χ for functional products. Exploiting the localized nature of $\mathbf{H}$ and $\mathbf{G}$ and using the asymptotic forms (3.2) we find to leading order in ϵ [8,9]

$$(3.5) \qquad \begin{aligned} \dot{\chi}_j &= a_R e^{-\eta_R(\chi_j - \chi_{j+1})} \cos[\nu_R(\chi_j - \chi_{j+1}) + \psi_R] \\ &+ a_L e^{-\eta_L(\chi_{j-1} - \chi_j)} \cos[\nu_L(\chi_{j-1} - \chi_j) + \psi_L], \end{aligned}$$

where a_R, a_L, ψ_R and ψ_L are parameters that can be calculated for a specific realization of the reaction-diffusion system (2.1) and $\chi_j > \chi_i$ for $j < i$.

Equations like (3.5) describe the time evolution of widely spaced wavetrains of impulses. The velocity of a given impulse is determined by the local field-values of nearby impulses. These values are exponentially small, yet significant because of the marginal stability of the solitary-wave solution to translations. The present theory is particularly applicable to excitable media where impulse collisions are naturally avoided [5]. In such media the rate of growth, η_R, is normally much larger than the rate of decay, η_L. Consequently, when the distribution of impulse spacings is sufficiently narrow, the first term on the rhs of (3.5) can be neglected and the impulse equations simplify to

$$(3.6) \qquad \dot{\chi}_j = H_L(\chi_{j-1} - \chi_j) \; H_L(\lambda) := a_L e^{-\eta_L \lambda} \cos(\nu_L \lambda + \psi_L).$$

The asymptotic form (3.2b) with $\nu_L \neq 0$ describes a medium with (damped) oscillatory recovery. We will show now that this manner of recovery gives rise to a form of spatial complexity. Consider a train of N impulses. The first impulse is stationary in a frame moving with speed c_0, for there is no impulse ahead of it to perturb its motion. Thus $\dot{\chi}_1 = 0$. All other impulses ($j = 2, ..., N$) are affected by the recovery phase of the medium just ahead of them and satisfy (3.6). Steady state solutions of (3.6) ($\dot{\chi}_j = 0$) describe trains of impulses that propagate with uniform speed c_0. There are many ways to construct such solutions, for any configuration of impulses with spacings $\lambda_j := \chi_{j-1} - \chi_j$ such that $H_L(\lambda_j) = 0$ is a solution. The criterion for stability, $H'_L(\lambda_j) > 0$ for any spacing in the train, still leaves us with many solutions. We therefore conclude that a multitude of stable wavetrains, mostly nonuniform, can exist in the case of oscillatory recovery. As the number of impulses, N, becomes large the systems develops extreme sensitivity to initial conditions and to external noise.

When the exponents η_R and η_L are of similar size and the head of the solitary wave is monotonic ($\nu_R = 0$) one can still understand asymptotic solutions of (3.5) rather easily. Setting $\dot{\chi}_j = const.$ in (3.5) we find a *pattern map* which relates a given spacing to the successive one [8,9]. Periodic wavetrains and finite wavetrains correspond, respectively, to periodic and homoclinic orbits of that map. The kind of spatial complexity discussed above pertains to a pattern map in a chaotic regime.

4. Interacting Vortices in Two Space Dimensions. The localized structures of one-dimensional systems generalize, in two dimensions, to wavefronts that are localized in one direction but extend uniformly in the other direction. Expanding ring patterns in excitable media provide an example. There is, however, a new kind of localized structure in two dimensions. In excitable media it corresponds to the free end of a terminated wavefront, while in oscillating media it appears as a point where the amplitude of oscillations vanishes. In both cases the patterns that surround the structures take the form of spiral waves or vortices [3,5,10]. The dynamics of a single spiral wave pose a number of interesting questions [1,11-14] which we will not dwell upon here. We will proceed directly to the many body problem of interacting vortices and consider primarily the more tractable case of oscillating media.

Suppose that at a given value of some control parameter the rde's (2.1) undergo a supercritical Hopf bifurcation. The equilibrium state, $\mathbf{U} = \mathbf{0}$, becomes unstable and a new, oscillatory state emerges:

(4.1) $$\mathbf{U}(\mathbf{r}, t) = Z(\mathbf{r}, t) \exp(i\omega_0 t) \mathbf{U}_0 + c.c..$$

Here, Z is a complex amplitude that depends weakly on space and time, ω_0 is the frequency of a uniformly oscillating medium and $\mathbf{U}_0$ is a constant vector. Close enough to the onset of oscillations the amplitude Z satisfies the equation [3]

(4.2) $$\partial_t Z = \mu Z - (1 + i\alpha)|Z|^2 Z + (1 + i\beta)\nabla^2 Z,$$

where μ, α and β are real constants. Equation (4.2) has spiral-wave solutions of the form

(4.3) $$Z_v(\mathbf{r}, t) = z(\mathbf{r})e^{i\omega t}, \quad z(\mathbf{r}) = A(r)e^{in\theta + is(r)},$$

where r and θ are polar coordinates in the plane, n is an integer and $A(r)$ and $s(r)$ have the asymptotic forms [3,15]:

(4.4)
$$A(r) \sim r^{|n|}, \ s(r) \sim r^{|n|+1} \ r \to 0$$
$$A(r) \sim A_0 + c/r, \ s(r) \sim pr + q\ln r \ r \to \infty.$$

Here, $\omega = -\alpha\mu + (\alpha - \beta)p^2$, $A_0 = \sqrt{(\mu - p^2)}$, $c = p(1+\beta^2)/[2A_0(\alpha - \beta)]$, $q = -(1 + \alpha\beta)/[2(\alpha - \beta)]$, and $p = p(\mu, \alpha, \beta)$ is the selected asymptotic wavenumber for an isolated vortex [15,16]. The integer n in (4.3) is given by the integral $(2\pi)^{-1} \oint \nabla\varphi \cdot d\mathbf{r}$ over a closed loop that contains the spiral core, where $\varphi = \arg(z)$, and is known as the topological charge or the winding number of the spiral wave (vortex).

In section 3 we used the solitary-wave solution as the basic building block for construction of extended wavetrains. Here, we will use the single vortex solution (4.3) to construct multi-vortex states. As before, we utilize the continuous symmetries of the system, this time the symmetries of (4.2): *space translations*, $Z(\mathbf{r}) \to Z(\mathbf{r} - \mathbf{r}_0)$, and *rotations* in the complex Z plane, $Z \to Z\exp(i\psi)$. Acting with these symmetries on (4.3) we find another solution of (4.2), $z(\mathbf{r} - \mathbf{r}_0)\exp(i\omega t + i\psi)$. A multi-vortex solution is now approximated by

(4.5)
$$Z_{mv}(\mathbf{r}, t) = \frac{e^{i\omega t + i\psi}}{A_0^{N-1}} \prod_{j=1}^{N} z_j(\mathbf{r} - \mathbf{r}_j),$$

where $z_j(\mathbf{r})$ is given by (4.3) with winding number n_j. If we allow the symmetry parameters $\mathbf{r}_j$ to depend on time and ψ to depend on space and time we find solvability conditions, in the form of evolution equations, which guarantee that (4.5) is indeed an approximate solution of (4.2). To write down the evolution equations we find it useful to define a "center of mass" coordinate

$$\mathbf{R}_{cm} = \sum_{j=1}^{N} |n_j| \mathbf{r}_j \ \bigg/ \ \sum_{j=1}^{N} |n_j|,$$

and a new phase

$$\phi(\mathbf{r} - \mathbf{R}_{cm}, t) = \psi(\mathbf{r}, t) + (N-1)p|\mathbf{r} - \mathbf{R}_{cm}|.$$

In terms of these quantities the evolution equations read

(4.6)
$$\frac{|n_i|}{2}\dot{\mathbf{r}}_i = \beta|n_i|\nabla\phi_{|\mathbf{r}_i} + n_i\nabla\phi_{|\mathbf{r}_i} \times \hat{\mathbf{k}} + \sum_{j\neq i}(n_in_j + \beta|n_i|q)\frac{\mathbf{r}_i - \mathbf{r}_j}{|\mathbf{r}_i - \mathbf{r}_j|^2}$$
$$- \sum_{j\neq i}(\beta n_j|n_i| - n_iq)\frac{(\mathbf{r}_i - \mathbf{r}_j) \times \hat{\mathbf{k}}}{|\mathbf{r}_i - \mathbf{r}_j|^2},$$

(4.7)
$$\partial_t\phi - \dot{\mathbf{R}}_{cm} \cdot \nabla\phi = 2p(\alpha - \beta)\partial_{rcm}\phi + (1 + \alpha\beta)\nabla^2\phi + (\alpha - \beta)(\nabla\phi)^2$$
$$- 2p^2A_0^{-2}(1 + \alpha^2)\partial_{rcm}^2\phi + pA_0^{-2}(\alpha - \beta)(1 - \alpha\beta)\nabla^2\partial_{rcm}\phi$$
$$- pA_0^{-2}(\alpha + \beta)(1 + \alpha\beta)\partial_{rcm}\nabla^2\phi - \frac{1}{2}A_0^{-2}\beta^2(1 + \alpha^2)\nabla^4\phi + h.o.t..$$

Here, $\hat{\mathbf{k}}$ is a unit vector perpendicular to the plane and $r_{cm} \equiv |\mathbf{r} - \mathbf{R}_{cm}|$. In deriving equations (4.6) and (4.7) we assumed that p is a small parameter ($p^2 \ll \mu$) and that $\nabla\phi$ and $|\mathbf{r}_i - \mathbf{r}_j|^{-1}$ are of $O(p)$. In physical terms, we demand that the inter-vortex distance and the asymptotic wavelength are both much larger than the size of the vortex core. Thus (4.7) includes terms to $O(p^4)$. For a vortex-free and uniform medium ($p = 0$) equation (4.7) reduces to the Kuramoto-Sivashinsky equation [3].

The requirement that the coefficient of $\partial^2_{r_{cm}} \phi$ vanishes gives an Eckhaus instability relation [3,17]

$$(4.8) \qquad p^2 = \mu(1 + \alpha\beta)/(3 + \alpha\beta + 2\alpha^2).$$

This relation together with the form, $p = p(\mu, \alpha, \beta)$, of the asymptotic wavenumber of the spiral solution (4.3), define a range of parameters in which the truncated form of (4.7),

$$(4.9) \qquad \partial_t\phi - \dot{\mathbf{R}}_{cm} \cdot \nabla\phi = 2p(\alpha - \beta)\partial_{r_{cm}}\phi + (1 + \alpha\beta)\nabla^2\phi + (\alpha - \beta)(\nabla\phi)^2,$$

is expected to be valid.

Equations (4.6) and (4.9) can be used to derive the interaction between a pair of vortices. We will consider here the simpler case of $n_1 = n_2 = 1$ and look for axisymetric solutions, $\phi = \phi(r_{cm})$, of (4.9). It is straightforward to show that in this case $\dot{\mathbf{R}}_{cm} = 0$. Equation (4.9) simplifies then considerably and can be solved by a Hopf-Cole transformation. We obtain

$$(4.10) \qquad \phi = \phi_\ell = \Omega_\ell t - \pi_\ell r_{cm} - 2q \ln[P_\ell(a_\ell r_{cm})],$$

where

$$\Omega_\ell = -p^2(\alpha - \beta)(1 - \lambda_\ell^2), \quad \pi_\ell = p(1 - \lambda_\ell) \quad \text{and} \quad a_\ell = pq^{-1}\lambda_\ell$$

with $\lambda_\ell = (2\ell + 1)^{-1}$ ($\ell \in \mathcal{Z}^+$). $P_\ell(x)$ is a polynomial of degree ℓ that solves the linear differential equation

$$(4.11) \qquad xP_\ell'' + (1 - x)P_\ell' + \ell P_\ell = 0.$$

Equation (4.11) can easily be solved for any degree ℓ. For $x \gg 1$ the solution simplifies to $P_\ell(x) \approx x^\ell$. Such a condition is attainable when considering large inter-vortex distances (compared with p^{-1}) or when $q \propto (1 + \alpha\beta)$ becomes small (i.e. close to the Eckhaus instability curve). Using (4.10) in (4.6) we then find

$$(4.12a) \qquad \frac{1}{4}\dot{\rho} = -\beta\pi_\ell + [1 - \beta q(4\ell - 1)]\frac{1}{\rho},$$

$$(4.12b) \qquad \frac{\rho}{4}\dot{\vartheta} = \pi_\ell + [\beta + q(4\ell - 1)]\frac{1}{\rho},$$

where $\rho \equiv |\mathbf{r}_2 - \mathbf{r}_1|$ is the inter-vortex distance and ϑ is the angle between $\mathbf{r}_2 - \mathbf{r}_1$ and a fixed direction in the plane.

Equations (4.12) describe the dynamics of a vortex pair with winding numbers $n_1 = n_2 = 1$. Stable steady state solutions of (4.12a) correspond to bound pairs of vortices. For $\beta \neq 0$ such solutions do exist. The inter-vortex distances of these pairs are given by

$$(4.13) \qquad \rho_\ell^* = \frac{1 - \beta q(4\ell - 1)}{\beta \pi \ell},$$

while the condition for stability is $\beta p > 0$. When βq is positive only a few bound-pair states exist. They correspond to inter-vortex distances ρ_ℓ^* with $\ell = 1, ..., \ell_{max}$, where ℓ_{max} is the integer part of $(1 + \beta q)/4\beta q$. For βq negative a whole family of bound-pair states exists (that is, ℓ can be any positive integer). Another consequence of (4.13) is that bound pairs of vortices (with equal winding numbers) rotate with constant angular velocity, $\dot{\vartheta}(\rho_\ell^*)$.

The bound-pair solutions disappear when $\beta \to 0$ despite the fact that vortices with spiral structures still exist in that limit. The parameter β is a measure of the size of phase-field contributions to vortex interactions. For small β values the phase-field effect becomes significant only at large inter-vortex distances where the $1/\rho$ repulsion term is sufficiently small. In that case only vortices whose distances are large enough (of order $(\beta p)^{-1}$) can form bound pairs. At relatively large β values the phase field is already effective at distances of $O(p^{-1})$.

The fact that vortices in oscillating media can form bound pairs should be attributed to the mediating effect of the spiral phase field; vortices with rectilinear equiphases either attract or repel each other. A somewhat similar situation applies to excitable media. Recent numerical simulations [18] indicate that bound vortex-pairs may exist in this kind of medium as well. For winding numbers $n_1 = n_2 = 1$ the vortices rotate around a fixed center while for $n_1 = -n_2 = 1$ they drift in a direction normal to $\mathbf{r}_1 - \mathbf{r}_2$ (drifting vortex pairs have also been found in numerical simulations on (4.2) [19]). The difference between oscillating and excitable media seems to lie in the nature of the mediating spiral field. In the former it is an *extended* sinusoidal field while in the latter it resembles an array of *localized* solitary wave-fronts [14]. Consequently vortices in excitable media which are a few wavelengths apart appear independent [18]; the interaction between the two cores is screened by the surrounding spiral wavefronts.

5. Discussion. We have presented here an effort to understand extended patterns in reactive media in terms of their localized constituents: solitary waves in one dimension, vortices in two dimensions. Evolution equations for these structures have been derived by exploiting the continuous symmetries of the system. The equations are valid for sparse patterns where typical distances between nearby structures are considerably larger than their widths. The approach pursued here does not allow for annihilation and nucleation of structures and therefore does not apply, in its present form, to turbulent patterns where such processes are important. Yet, results obtained with this approach may bear on the transition to turbulence, as we now discuss.

Recent numerical simulations on (4.2) suggest that the transition to turbulence in oscillating media may occur by the spontaneous nucleation of vortices [20]. Above the transition, vortices are continuously being created and annihilated so as to keep a steady mean number. The result that vortices can form bound pairs may have the interesting consequence of making the transition hysteretic. Suppose we prepare a system to contain only one vortex and drive it to the turbulent regime. This can be done in a number of ways, but we choose to consider here the path in parameter space, $\alpha = -\beta < 0$, for which numerical data on the selected asymptotic wavenumber, $p = p(\mu, \alpha, \beta)$, are available [16]. Upon increasing β, a critical value β_c^+ is reached where phase instabilities develop. This value, determined by (4.8) and the asymptotic wavenumber p, marks the onset of turbulence since phase instabilities seem to provide the driving force needed for the nucleation of additional vortices [20]. On going backward below β_c^+, vortices which are sufficiently apart from each other to form bound pairs, will not annihilate each other and a multi-vortex state will result. On further decreasing β, the minimal distance to form bound pairs will increase because it is proportional to $(\beta p)^{-1}$ and p decreases with $\beta = -\alpha$ [16,19]. As a result more and more annihilation events will take place until a value $\beta_c^- < \beta_c^+$ is reached where the minimal distance exceeds the size of the system and the initial one-vortex state is recovered.

Vortex-initiated turbulence appears to be a general phenomenon in two-dimensional nonequilibrium systems; similar transitions have recently been observed in electroconvecting nematic liquid crystals [21] and in capillary ripples [22]. It would be of interest to know whether this type of transition may occur in excitable media as well.

REFERENCES

[1] A. T. Winfree, *When Time Breaks Down*, Princeton, New Jersey, 1987.

[2] J. J. Tyson and J. Keener, *Singular perturbation theory of traveling waves in excitable media (a review)*, Physica D, 32 (1988), pp. 327-361.

[3] Y. Kuramoto, *Chemical Oscillations, Waves, and Turbulence*, Springer, Berlin, 1984.

[4] A. L. Hodgkin and A. F. Huxley, *A quantitative description of membrane current and its application to conduction and excitation in nerve*, J. Physiol., 117 (1952), pp. 500-544. Springer-Verlag, Berlin, 1984.

[5] A. T. Winfree, *Spiral waves of chemical activity*, Science 175 (1972), pp. 634-636.

[6] B. Friedman, *Principles and Techniques of Applied Mathematics*, Chapman and Hall Ltd., London, 1956.

[7] C. Elphick and E. Meron, *Localized structures in surface waves*, Phys. Rev. A, 40 (1989), pp. 3226-3230

[8] C. Elphick, E. Meron and E. A. Spiegel, *Patterns of propagating pulses*, to appear in SIAM J. Appl. Math..

[9] C. Elphick, E. Meron and E. A. Spiegel, *Spatiotemporal complexity in traveling patterns*, Phys. Rev. Lett., 61 (1988), pp. 496-499.

[10] The manner by which a rectilinear wavefront with a free end evolves toward a rotating spiral wave has been studied in E. Meron and P. Pelcé, *Model for spiral wave formation in excitable media*, Phys. Rev. Lett., 60 (1988), pp. 1880-1883.

[11] W. Jahnke, W. E. Skaggs, and A. T. Winfree, *Chemical vortex dynamics in the Belousov-Zhabotinsky reaction and in the two- variable Oregonator Model*, J. Phys. Chem., 93 (1989), pp. 740-749, and references therein.

[12] S. C. Muller and B. Hess, *Nonlinear dynamics in chemical systems*, in *Cooperative Dynamics in Complex Physical Systems*, Ed. H. Takayama, Springer, Berlin, 1989.

[13] K. I. Agladze, V. A. Davydov, and A. S. Mikhailov, *Observation of a helical-wave resonance in an excitable distributed medium*, Pis'ma Zh. Eksp. Teor. Fiz., 45 (1987) pp. 601-603.

[14] E. Meron, *Nonlocal effects in spiral waves*, Phys. Rev. Lett., 63 (1989), pp. 684-687.

[15] P. S. Hagan, *Spiral waves in reaction-diffusion equations*, SIAM J. Appl. Math., 42 (1982), pp. 762-786.

[16] E. Bodenschatz, *Muster und defekte im rahmen der schwach nichtlinearen analyse von anisotropen structurbildenden systemen*, Ph.D. Thesis, University of Bayreuth, 1989.

[17] B. A. Malomed, *Nonlinear waves in nonequilibrium systems of the oscillatory type, part 1*, Z. Phys. B, 55 (1984), 241-248.

[18] E. A. Ermakova, A. M. Pertsov, and E. E. Shnol, *On the interaction of vortices in two-dimensional active media*, Physica D, 40 (1989), 185-195.

[19] E. Bodenschatz, M. Kaiser, L. Kramer, W. Pesch, A. Weber and W. Zimmermann, *Patterns and defects in liquid crystals*, to appear in *New Trends in Nonlinear Dynamics and Pattern Forming Phenomena: The Geometry of Nonequilibrium*, Eds. P. Coullet and P. Huerre, Plenum Press, 1989.

[20] P. Coullet, L. Gil, and J. Lega, *Defect-mediated turbulence*, Phys. Rev. Lett., 62 (1989), 1619-1622.

[21] I. Rehberg, Steffen Rasenat, and Victor Steinberg, *waves and defect-initiated turbulence in electroconvecting nematics*, Phys. Rev. Lett., 62 (1989), pp. 756-759.

[22] N. B. Tufillaro, R. Ramshankar, and J. P. Gollub, *Order-disorder transition in capillary ripples*, Phys. Rev. Lett., 62 (1989), pp. 422-425.

SOME RECENT RESULTS IN CHEMICAL REACTION NETWORK THEORY

MARTIN FEINBERG*

Abstract. The aim of chemical reaction network theory is to draw connections between reaction network structure and qualitative properties of the corresponding differential equations.
Some recent results are discussed, in particular those relating to the possibility of multiple steady
states in very complex continuous flow stirred tank reactors, to mechanism discrimination in heterogeneous catalysis, and to the possibility of traveling composition waves on isothermal catalyst
surfaces

Key words. chemical reaction networks, ordinary differential equations, reaction-diffusion
equations, traveling waves, Conley index, SCL graphs, catalysis, CFSTR.

1. Introduction. Chemists and engineers are required to confront a bewilder-
ing array of distinct chemical systems, each with its own set of molecular species
and its own network of chemical reactions. In a real reactor, the amounts of the
various species will generally be governed by a large first order system of differential
equations, almost always nonlinear (and usually polynomial). The equations often
contain many parameters (e.g., rate constants) which are known only roughly if
they are known at all. To make matters worse, each new chemical system gives rise
to a completely new system of differential equations.

At first glance, the situation would seem hopeless. Despite great progress in
dynamical systems, it remains quite difficult to study a first order system of polynomial differential equations in, say, seven or eight dependent variables. It is worth
remembering that the Lorenz system, with all the complexity it carries, is a polynomial system in only three variables and that Hilbert's Sixteenth Problem is about
the behavior of polynomial equations in only two variables.

The fact is, however, that the differential equations that arise in the study of
chemical reactors have a very special structure. Given a reactor of specified type–for
example, the classical continuous flow stirred tank reactor–undergraduate students
in engineering know (or should know) how to pass from an indicated network of
chemical reactions to the appropriate system of differential equations. In fact, when
reaction rates are governed by the usual mass action kinetics, the polynomial vector
field is completely determined (up values of parameters such as rate constants)
by the structure of the underlying reaction network. It is this precise connection
between reaction network structure and the shape of the governing equations that
gives the mathematics of chemical reactors its special character.

The central question of chemical reaction network theory can be stated in the
following way: *What is the relationship between the structure of a reaction network
and the variety of phase portraits that the corresponding differential equations might*

*Department of Chemical Engineering, University of Rochester, Rochester, NY, USA 14627.
Work described in this article was supported by National Science Foundation grants to Martin
Feinberg and (for the research on traveling waves) to David Terman.

exhibit (as parameter values vary)? That is, for a specified network, one would like to have practical means to determine whether there are rate constant values such that the corresponding differential equations admit Phenomenon X. Here Phenomenon X might be the existence of multiple rest points, of periodic orbits, and so on. By a "practical" theory I mean one that can be implemented (perhaps with the help of a computer) by a chemist or engineer without much training in mathematics.

Although this won't be a "from scratch" tutorial on reaction network theory, it is intended as much for mathematicians as for chemists and engineers. In any case, I hope I can convey a sense of what the problems are and of what some results look like. Mathematicians who are encountering chemical reactors for the first time will be able to get a rough feeling for the way differential equations are formulated, but they might want to consult [5] for a fuller introduction. The same reference contains a review of results in chemical reaction network theory as they stood a decade ago. References [6] and [7] cover a lot of the same ground, but from a perspective informed by more recent theory. Although they too are meant to be expository, it is probably better to read [5] first.

Two presumptions will be in force for all the systems considered here. First, the temperature will be understood to be fixed and spatially uniform. In this case, complexity in the differential equations will derive entirely from the interplay of the various reactions and not at all from thermal effects. Second, all reaction rates will be governed by mass action kinetics. This means, for example, that if c_A and c_B are the local molar concentrations of species A and B, then the local occurrence rate per unit volume of a reaction such as $A \rightarrow Q + R$ should be proportional to c_A while the rate of a reaction such as $A + B \rightarrow S + T$ should be proportional to $c_A c_B$. (The proportionality constants are the *rate constants* for the reactions.) The rough idea here is that rate of the first reaction should depend simply on the amount of A present while the occurrence rate of the second reaction should reflect the likelihood of a collision between A and B.

I'll describe results in three areas: (i) the relationship between reaction network structure and the possibility of multiple rest points in very complex continuous flow stirred tank reactors, (ii) the problem of mechanism discrimination in heterogeneous catalysis, and, finally, (iii) the extent to which the classical mechanisms for heterogeneous catalysis induce traveling composition waves on catalytic surfaces. Without much preparatory work, I can penetrate fairly deeply into the first of these areas, and so I am going to concentrate my efforts there. Discussion of the two remaining areas will be more abbreviated.

However, I do want to mention in this introductory section that Conley Index methods (in the hands of Dave Terman) are crucial to the results on traveling waves. Charlie Conley was very enthusiastic about chemical reaction network theory, and, for me, his enthusiasm was a source of encouragement when encouragement was very much needed. At the time, I did not know that Charlie was enthusiastic about everything.

2. Multiple rest points in homogeneous continuous flow stirred tank reactors with complicated chemistry. A *continuous flow stirred tank reactor* (CFSTR) is a vessel to which reactants are continuously added (in a feed stream) and from which the reacting mixture is continuously withdrawn (in an effluent stream). I shall suppose that the mixture in the vessel is stirred so zealously that it is homogeneous at each instant. I'll also suppose that the composition of the feed stream is fixed, that the volumetric flow rates of the feed and effluent streams are constant and identical, and that the volume of the mixture in the vessel remains constant. The *residence time* of the reactor, which I'll denote by θ , is the mixture volume divided by the volumetric flow rate.

I'll begin by considering a CFSTR in which there are five species, A, B, C, E and F, and I'll suppose that the reactions among the species are those shown (2.1). The

$$A+B \xrightarrow{1} E$$
(2.1)
$$B+C \xrightarrow{2} F$$
$$C \xrightarrow{3} 2A$$

species concentrations in the feed stream are denoted $c_A^f, c_B^f, c_C^f, c_E^f$ and c_F^f. (Here the concentrations are in moles per volume.) The species concentrations in the vessel are, at each instant, identical to those in the effluent stream, but they vary with time by virtue of the occurrence of chemical reactions. These concentrations I'll denote by c_A, c_B, c_C, c_E and c_F.

The governing differential equations are shown in (2.2). Here k_1, k_2 and k_3 are rate constants for the corresponding reactions in (2.1). Terms in which the rate constants appear account for contributions of the various chemical reactions to changes in the species concentrations within the vessel. Terms containing the residence time account for contributions of the feed and effluent streams.

(2.2)
$$
\begin{aligned}
c_A' &= (1/\theta)(c_A^f - c_A) - k_1 c_A c_B + 2k_3 c_C \\
c_B' &= (1/\theta)(c_B^f - c_B) - k_1 c_A c_B - k_2 c_B c_C \\
c_C' &= (1/\theta)(c_C^f - c_C) - k_2 c_B c_C - k_3 c_C \\
c_E' &= (1/\theta)(c_E^f - c_E) + k_1 c_A c_B \\
c_F' &= (1/\theta)(c_F^f - c_F) + k_2 c_B c_C
\end{aligned}
$$

It is already apparent that even simple reaction networks can lead to dynamical equations that are not so easy to study. Different networks lead to different CFSTR equations. Our interest is in relationship between qualitative properties of the equations and the structure of the reaction network from which they came.

There are many questions that can be asked, but in this section I'll be interested in only one: *What is the relationship between reaction network structure and the possibility of multiple positive rest points in the corresponding homogeneous CFSTR equations?* By a positive rest point I mean one in which all the species concentrations are positive. In this section, *when I say that a reaction network has the capacity for multiple positive rest points, I'll mean that there is some assignment of a (positive) residence time, of (positive) rate constants, and of (non-negative) feed concentrations such that the corresponding homogeneous CFSTR equations admit at least two positive rest points.*

It happens that reaction network (2.1) has the capacity for multiple positive rest points. On the other hand, networks (2.3) and (2.4) do not, so it is already apparent that there is some subtlety here. At first glance, it would seem extremely difficult

$$A + B \rightarrow E$$

(2.3)
$$B + C \rightarrow F$$

$$C \rightarrow A$$

$$A + B \rightarrow E$$

$$B + C \rightarrow F$$

(2.4)
$$C + D \rightarrow G$$

$$D \rightarrow 2A$$

to determine whether a more elaborate network such as (2.5) has the capacity for multiple rest points. So that the full weight of the problem might be felt, I have displayed in (2.6) the CFSTR equations that correspond to network (2.5).

$$A + B \underset{2}{\overset{1}{\rightleftarrows}} D \underset{4}{\overset{3}{\rightleftarrows}} 2C$$

$$C \underset{6}{\overset{5}{\rightleftarrows}} E$$

$$A + F \underset{8}{\overset{7}{\rightleftarrows}} G$$

(2.5)
$$2B \underset{10}{\overset{9}{\rightleftarrows}} H$$

$$I$$

$$I + J \underset{16}{\overset{15}{\rightleftarrows}} K$$

$$c'_A = (1/\theta)(c_A^f - c_A) - k_1 c_A c_B + k_2 c_D - k_7 c_A c_F + k_8 c_G$$

$$c'_B = (1/\theta)(c_B^f - c_B) - k_1 c_A c_B + k_2 c_D - 2(k_9 + k_{14})c_B^2 + 2k_{10}c_H + 2k_{13}c_I$$

$$c'_C = (1/\theta)(c_C^f - c_C) + 2k_3 c_D - 2k_4 c_C^2 - k_5 c_C + k_6 c_E$$

$$c'_D = (1/\theta)(c_D^f - c_D) + k_1 c_A c_B + k_4 c_C^2 - (k_2 + k_3)c_D$$

$$c'_E = (1/\theta)(c_E^f - c_E) + k_5 c_C - k_6 c_E$$

(2.6) $\quad c'_F = (1/\theta)(c_F^f - c_F) - k_7 c_A c_F + k_8 c_G$

$$c'_G = (1/\theta)(c_G^f - c_G) + k_7 c_A c_F - k_8 c_G$$

$$c'_H = (1/\theta)(c_H^f - c_H) + k_9 c_B^2 + k_{12}c_I - (k_{10} + k_{11})c_H$$

$$c'_I = (1/\theta)(c_I^f - c_I) + k_{14}c_B^2 + k_{11}c_H - (k_{13} + k_{12})c_I - k_{15}c_I c_J + k_{16}c_K$$

$$c'_J = (1/\theta)(c_J^f - c_J) - k_{15}c_I c_J + k_{16}c_K$$

$$c'_K = (1/\theta)(c_K^f - c_K) + k_{15}c_I c_J - k_{16}c_K$$

I want to tell you how to determine – within about five minutes! – whether network (2.5) has the capacity to generate multiple rest points. The procedure is not only simple but also sufficiently delicate as to indicate why, despite the fact that network (2.1) can give multiple rest points, the slightly different networks (2.3) and (2.4) cannot.

The theory I'll describe results from joint work with Paul Schlosser, and I am happy to say that he was the real hero of effort. Proofs and additional results can be found in Schlosser's PhD thesis [11]. Reference [12] contains a fuller discussion and more examples. Some earlier work (leading to narrower results) is described in [10]. It should be kept in mind that discussion in this section is specific to homogeneous CFSTRs. When I say that such-and-such a network has the capacity for multiple positive rest points, I mean that we were able to construct rate constants, a residence time, a feed composition and two positive rest points which, when taken together, satisfy the appropriate homogeneous CFSTR equations.

I am going to tell you how to take a reaction diagram such as (2.5) and construct from it another diagram, called the *SCL Graph*. Then I'll state two theorems that relate the structure of the SCL Graph to the possibility of multiple positive rest points.

To begin, I want to introduce a small amount of vocabulary. Network (2.7) will help illustrate the few terms I'll need. The *complexes* of a reaction network are the

$$A + B \to C$$

(2.7)
$$
\begin{array}{ccc}
A + D & \rightleftarrows & E \\
& \searrow \quad \nearrow & \\
& 2F &
\end{array}
$$

objects that live at the heads and tails of the reaction arrows. The set of complexes for network (2.7) is indicated in (2.8). Note that in (2.7) each complex appears

(2.8) $$\{A + B, C, A + D, E, 2F\}$$

precisely once and that arrows are drawn to indicate a "reacts to" relation in the set of complexes. Such a display is called a *standard reaction diagram*. The standard reaction diagram induces a partition of the complex set into *linkage classes*: Note that the diagram (2.7) is made up of two distinct connected components, one containing the complexes $A + B$ and C, the other containing the complexes $A + D, E$ and $2F$. The linkage classes of a network are just the subsets of complexes that reside within the individual connected components that make up the standard reaction diagram. The linkage classes of network (2.7) are shown in (2.9). The way in which linkage classes are denoted has no real significance, but my practice will be to number them from top to bottom as they appear in whatever standard reaction diagram I am considering.

(2.9) $$L_1 = \{A + B, C\} \quad L_2 = \{A + D, E, 2F\}$$

Now I can tell you how the *Species-Complex-Linkage Graph* (SCL Graph) for a network is constructed. Again, I will use network (2.7) as an example. Its SCL Graph is shown in Figure 2.1. You begin by writing on paper a symbol for each species in the network and also a symbol for each linkage class. (After a while you learn how to arrange these things in a nice way.) Then, for each species, you do the following: If the species appears in a complex within a particular linkage class, you draw an arc connecting the species to the symbol for that linkage class, and you label the arc with the complex name. (If the species appears in several complexes within a particular linkage class, you draw an arc for each such complex.) For example, species A appears in linkage class L_1 (in complex $A + B$), so an arc (labeled with the complex name $A + B$) is drawn between A and L_1. Species A also appears in linkage class L_2 (in complex $A + D$), so an arc (labeled $A + D$) is drawn between A and L_2. After proceeding in this way through the entire species set, you obtain the SCL Graph shown in Figure 2.1.

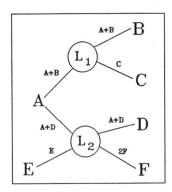

Figure 2.1. The SCL Graph for Network 2.7

It will be instructive to consider another example. The SCL Graph for network (2.10) is shown in Figure 2.2. The linkage classes for the network are

$$L_1 = \{A + B, C, A + D\}, \quad L_2 = \{2A, E + G\} \quad \text{and} \quad L_3 = \{A + E, F\}.$$

Note that species A appears in two complexes within linkage class L_1, and so there are two arcs (labeled $A + B$ and $A + D$) that connect A to L_1.

$$A + B \rightarrow C \rightleftarrows A + D$$

(2.10)
$$2A \rightleftarrows E + G$$

$$A + E \rightleftarrows F$$

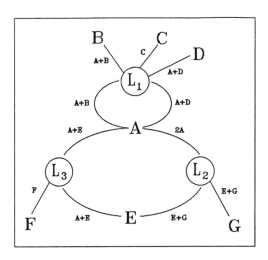

Figure 2.2.   The SCL Graph for Network 2.10

There is an important distinction between the SCL Graphs shown in Figures 2.1 and 2.2. The second contains two closed loops ($L_1 - A - L_1$ and $A - L_2 - E - L_3 - A$), but the first contains none. The presence of loops and their nature bears heavily on the capacity for multiple positive rest points. In fact, I am already in a position to state a theorem.

THEOREM 2.1. *Consider a chemical reaction network for which the SCL Graph contains no loops. If the network is governed by mass action kinetics, then the corresponding isothermal homogeneous CFSTR equations cannot admit multiple positive rest points, no matter what the feed composition might be and no matter what (positive) values the residence time and rate constants take.*

I promised that you would be able to determine very quickly whether the intricate network (2.5) has the capacity for multiple positive rest points. (Remember the daunting system of equations shown in (2.6).) The SCL Graph for network (2.5) is displayed in Figure 2.3. *It has no loops, so the system (2.6) cannot admit multiple positive rest points.* That's that.

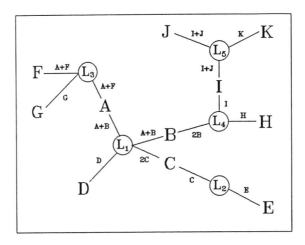

Figure 2.3. The SCL Graph for Network 2.5

What if the SCL Graph does have loops? In this case, there is still a lot that can be said, but I will need more vocabulary. In particular, I have to tell you about three kinds of loops: *o-loops, c-loops* and *s-loops*. These are not mutually exclusive; a loop in an SCL Graph can, for example, be both an *o*-loop and an *s*-loop.

By *c-pair* in an SCL Graph I mean a pair of (adjacent) arcs that carry the same complex label. (The letter "c" is supposed to remind you of the word complex.) In Figure 2.1, for example, the arcs $A - L_1$ and $L_1 - B$ constitute a *c*-pair because they both carry the complex label $A + B$. In Figure 2.2, the arcs $A - L_3$ and $L_3 - E$ both carry the complex label $A + E$, so they constitute a *c*-pair.

By an *o-loop* in an SCL Graph I mean a loop containing an odd number of *c*-pairs. (The letter "o" is supposed to remind you of the word odd.) In Figure 2.2

the loop $A - L_2 - E - L_3 - A$ is an o-loop because it contains precisely one c-pair (with arcs labeled $A+E$). The same loop contains only one arc of the c-pair labeled by complex $E + G$, so that c-pair is not counted in determining whether the loop is an o-loop. (Both arcs of the c-pair must lie in the loop.) In Figure 2.2 the loop $A - L_1 - A$ contains no c-pairs, so it is not an o-loop.

By a *c-loop* in an SCL Graph I mean a loop made up entirely of c-pairs. Consider, for example, network (2.11). Its SCL Graph is shown in Figure 2.4. (As I

$$A + B \rightleftarrows E$$
$$B + C \rightleftarrows F$$
(2.11)
$$C + D \rightleftarrows G$$
$$D + A \rightleftarrows H$$

indicated earlier, it will be my practice to number linkage classes from top to bottom as they appear in the standard reaction diagram under consideration.) The only loop in the figure is a c-loop because its set of arcs is the union of c-pairs. (On the other hand, the loop is not an o-loop because it contains four c-pairs.)

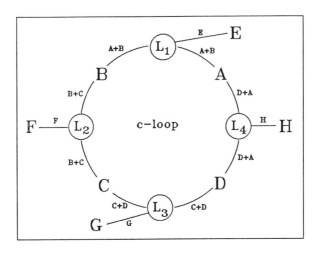

Figure 2.4.   The SCL Graph for Network 2.11

I will use network (2.12) to help explain what an *s-loop* is. The SCL Graph for (2.12) is displayed in Figure 2.5.

$$A + D \rightleftarrows 2B$$
(2.12)
$$B \rightleftarrows C$$
$$2C \rightleftarrows A + E$$

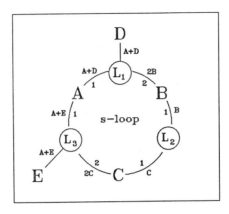

Figure 2.5. The SCL Graph for Network 2.12

To determine whether a loop is an s-loop, I shall first have to adorn its arcs with *stoichiometric coefficients*. When a species appears within a particular complex, the stoichiometric coefficient of a species within that complex is just the numerical multiplier of the species. For example, the stoichiometric coefficient of B in the complex $2B$ is 2. The stoichiometric coefficient of A in complex $A + D$ is 1. I adorn the arcs of a loop with stoichiometric coefficients in the following way: With each arc there is associated a species (at one end of the arc) and a complex (labeling the arc). Alongside each arc of the loop, I write the stoichiometric coefficient of the arc-end species within the labeling complex. (See Figure 2.5.)

Now I am going to do something strange. I am going to calculate a number for a loop by going around the loop, alternately multiplying and dividing the stoichiometric coefficients I encounter, until I arrive back at my starting point. For the loop in Figure 2.5, I will start at species A. Traversing the loop in a clockwise direction, I alternately multiply and divide the stoichiometric coefficients to make the calculation

$$1 \times (1/2) \times 1 \times (1/1) \times 2 \times (1/1) = 1.$$

I say that a loop is an s-loop if, as in the example, the result of this strange calculation is *one*. Thus, the loop shown in Figure 2.5 is an s-loop. (The "s" is supposed to remind you of stoichiometric coefficients.) To determine if a loop is an s-loop, the alternate multiplication and division of stoichiometric coefficients can begin anywhere in the loop. The direction of circumnavigation is immaterial.

Having introduced o-loops, c-loops and s-loops, I am almost in a position to state another theorem. I need just one more piece of terminology: A *loop arc* in an SCL Graph is an arc that is contained in a loop.

THEOREM 2.2. *Consider a reaction network such that in the SCL Graph*

(i) *each loop is an o-loop, a c-loop or an s-loop,*

(ii) *if both arcs of a c-pair are loop arcs, then no loop contains only one of the arcs,*

(iii) *if a loop is neither a c-loop nor an s-loop (in which case it is an o-loop), then no linkage class in that loop is adjacent to more than three loop arcs.*

If the network is governed by mass action kinetics, then the corresponding isothermal homogeneous CFSTR equations cannot admit multiple positive rest points, no matter what the feed composition might be and no matter what (positive) values the residence time and rate constants take.

REMARK. If there are no loops, all three conditions are satisfied trivially, so Theorem 2.2 subsumes Theorem 2.1. (Theorem 2.1 was just supposed to get you in the mood for Theorem 2.2.) Note that the "or" in (i) is not exclusive. That is, (i) requires only that each loop belong to at least one of the three categories described. Note also that (iii) is not a restriction on all o-loops. Rather, it is a condition that must be satisfied only by an o-loop which is neither a c-loop nor an s-loop.

We have studied only simple SCL Graphs, so it may be a little difficult to see how conditions (ii) and (iii) can fail. Let me consider two examples that are somewhat more complex.

Network (2.13) *does* have the capacity for multiple positive rest points, so its SCL Graph (shown in Figure 2.6) must fail to comply with one of the three conditions stated in Theorem 2.2. It is easy to see that every loop in Figure 2.6 is

$$A + B \rightleftarrows F$$

$$A + C \rightleftarrows G$$

(2.13)

$$C + D \rightleftarrows B$$

$$C + E \rightleftarrows D$$

an s-loop, whereupon (i) is satisfied and (iii) becomes vacuous. (This will be the situation for all networks such as (2.13) in which every stoichiometric coefficient is one.) To see that (ii) is not satisfied, note that the both arcs of the c-pair labeled by complex C + D are loop arcs. Note also that the loop $L_1 - A - L_2 - C - L_3 - B - L_1$ contains only one of the arcs.

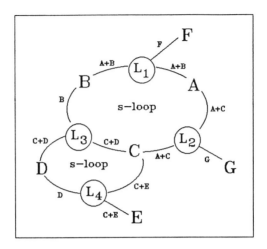

Figure 2.6.   The SCL Graph for Network 2.13

Network (2.14) also has the capacity for multiple positive rest points. Its SCL Graph is shown in Figure 2.7. Each loop contains just one c-pair, so both loops are o-loops, whereupon condition (i) is satisfied. It is easy to see that condition (ii) is satisfied as well. Thus, it must be condition (iii) that fails. In fact, the topmost loop is neither a c-loop nor an s-loop, and it contains a linkage class, L_2, that is adjacent to four loop arcs.

$$A + B \rightleftarrows E$$

(2.14)
$$C \rightarrow 2A \rightarrow F \leftarrow D \rightleftarrows B$$

$$C + D \rightleftarrows G$$

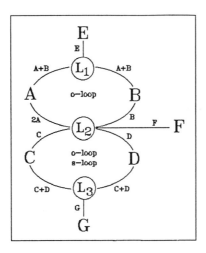

Figure 2.7. The SCL Graph for Network 2.14

Let me give another example. Consider network (2.15). There are 13 species and 18 reactions, so the corresponding CFSTR equations are even worse than those shown in (2.6). However horrible these equations might be, Theorem 2.2 tells us

$$A + B \rightleftarrows C$$
$$B + D \rightleftarrows E \rightleftarrows 2A$$
$$B + F \rightleftarrows G$$
$$(2.15) \qquad H + I \rightleftarrows J \rightleftarrows 2B$$
$$H + L \rightleftarrows K \rightleftarrows 2F$$
$$J + K \rightleftarrows M$$

very quickly that they cannot admit multiple rest points. The SCL graph for the network is drawn in Figure 2.8. There are four loops:

$$A - L_1 - B - L_2 - A, \qquad\qquad B - L_3 - F - L_5 - H - L_4 - B,$$
$$H - L_5 - K - L_6 - J - L_4 - H, \quad B - L_3 - F - L_5 - K - L_6 - J - L_4 - B.$$

The first is an o-loop, and the others are s-loops, so condition (i) in Theorem 2.2 is satisfied. It is easy to confirm that conditions (ii) and (iii) are satisfied as well. (In checking condition (iii), it might be noted that the topmost loop is neither an s-loop nor a c-loop and that the loop contains a *species* adjacent to four loop arcs. Condition (iii) requires only that no *linkage class* be adjacent to four loop arcs.)

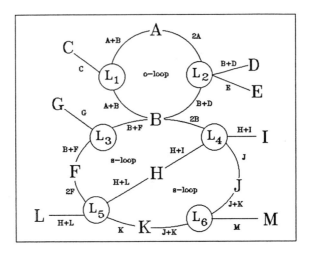

Figure 2.8. The SCL Graph for Network 2.15

Finally, I want to return to the very simple (but somewhat mysterious) examples with which we began. Remember that, among the very similar networks (2.1), (2.3) and (2.4), only (2.1) has the capacity for multiple positive rest points. The SCL Graphs for the networks are shown in Figures 2.9, 2.10 and 2.11. In each graph there is only one loop, and from this it easy to see that conditions (ii) and (iii) of Theorem 2.2 are satisfied for all three networks. It remains to determine for which of the networks condition (i) is satisfied. The loop in Figure 2.10 is an s-loop, and the loop in Figure 2.11 is an o-loop. Theorem 2.2 tells us, therefore, that neither network (2.3) nor (2.4) has the capacity for multiple positive rest points. In Figure 2.9, however, the loop is not an s-loop, an o-loop, or a c-loop, in which case Theorem 2.2 leaves the capacity of network (2.1) for multiple positive rest points undecided.

In fact, there are positive results in Paul Schlosser's thesis [11] which, unlike Theorems 2.1 and 2.2, indicate when a network *does* have the capacity for multiple rest points in a homogeneous CFSTR context. (We expect to make these results available in an article for Chemical Engineering Science [13].) I won't try to describe them here, except to say that they are not as wide-ranging as the results I've already stated.

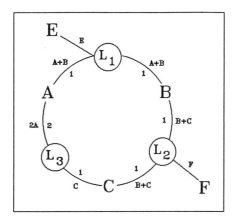

Figure 2.9.   The SCL Graph for Network 2.1

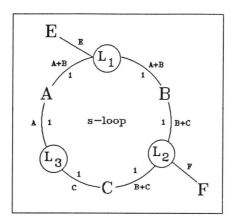

Figure 2.10.   The SCL Graph for Network 2.3

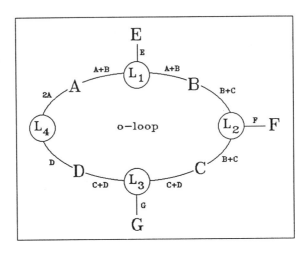

Figure 2.11. The SCL Graph for Network 2.4

I do, however, want to describe one interesting consequence of the positive results. Consider the family of networks (parameterized by the integer n) shown in (2.16).   Note that networks (2.1) and (2.3) are special members of the family,

$$A_0 + A_1 \rightarrow B_1$$
$$A_1 + A_2 \rightarrow B_2$$

(2.16)
$$\vdots$$

$$A_{n-1} + A_n \rightarrow B_n$$
$$A_n \rightarrow 2A_0$$

corresponding to $n = 2$ and $n = 3$. For each n, the induced SCL Graph has just one loop, just as in Figures 2.9 and 2.11. In each case, the loop is neither a c-loop nor an s-loop. The number of c-pairs in the loop is n, so the loop is an $o-loop$ precisely when n is odd. Thus, for odd n, Theorem 2.2 precludes the possibility of multiple positive rest points. On the other hand, for each even n – in particular for network (2.1) – the positive results indicate that there are multiple positive rest points for at least some values of the residence time, rate constants and feed concentrations. In other words, *networks in the family (2.16) have the capacity for multiple positive rest points if and only if n is even.*

I want to say a few words about what is seen in nature. There are certainly experiments that indicate the possibility of multiple rest points in isothermal homogeneous CFSTRs, but those experiments are fairly recent. In fact, it is almost doctrine among chemists and chemical engineers that there should be only one rest point, and it is common to hear reference to "the" steady state, as if uniqueness of rest points were somehow a consequence of natural law.

This view of things is understandable. It emerged from an acquired experience that is, I think, explained in part by Theorems 2.1 and 2.2. It is probably the case that reaction networks selected in some indiscriminate way satisfy the hypotheses of the theorems far more often than not. Among those networks that fail to comply with the hypotheses, there will be some that nevertheless have no capacity for multiple rest points. And the remaining networks will admit multiple rest points only for certain (perhaps rare) combinations of residence time, rate constants and feed composition. Even then, the basin of attraction for one of the rest points might be very large, in which case the other stable rest points would have low probabilities of detection.

Remember that the discussion in this section was tailored very specifically to homogeneous CFSTRs. I'll turn next to CFSTRs involving *heterogeneous catalysis*. For them, the situation is very different.

3. Mechanism discrimination in heterogeneous catalysis. I want to describe a reactor which, in its outward appearance, looks very much like the CFSTR studied in the last section. There, the tacit understanding was that reactions take place homogeneously throughout the fluid occupying the vessel. Here things will be different. Reactions will not take place in the fluid phase at all but, rather, on a solid catalytic surface in contact with the fluid phase. This interaction of solid and fluid phases is the defining characteristic of heterogeneous catalysis.

So that I can introduce the subject, consider once again a vessel to which reactants are continuously supplied (in a feed stream) and from which reactants and product are continuously withdrawn (in an effluent stream). In this case, think of the two streams as being gaseous rather than liquid. I'll suppose that, apart from an inert carrier such as argon, the feed stream contains only two species, A and B, and that the effluent stream contains the carrier gas, the reactants A and B, and also a product C.

I'll also suppose that all steady state experiments suggest the occurrence of the reaction

$$(3.1) \qquad\qquad A + B \to C.$$

By this I mean that, in all steady state experiments, the loss of a molecule of A between the feed and effluent streams is accompanied by the loss of a molecule of B and by the gain of a molecule of C. It would appear, then, that the chemical events are as simple as can be: In the gaseous mixture within the chamber, a molecule of A combines directly with a molecule of B to form a molecule of C.

But things are not always what they seem. Some of the most important chemical reactions are, in a sense, only apparent. They do not proceed directly, at least not at appreciable rates. Instead, they proceed by an indirect route mediated by a catalyst. The catalyst is often a metal such as platinum, palladium or rhodium.

I have only told you what is coming out of the vessel and what is going in. I haven't said very much about what is actually in the vessel, and now I want to suppose that the apparent reaction (3.1) is being promoted by a solid catalyst.

To keep the picture simple, imagine that the bottom of the vessel is made of a metallic catalyst and that a gaseous mixture of A, B and C sits above this catalytic surface. There is no appreciable reaction in the gas phase. Rather, C is produced from A and B through a mechanism of the following kind: One or both of the reactants bind from the gas phase onto active catalyst sites, surface-bound species react with each other (or with species in the adjacent gaseous mixture) to produce C, and C then desorbs from the catalyst to enter the gas phase. All the while, A and B are being introduced into the gas phase via the feed stream, and gaseous mixture is being withdrawn continuously in the effluent stream.

For the hypothetical reactor under consideration, I am going to suppose that, to good approximation, both the gas phase and the catalyst surface remain spatially homogeneous at all times. That is, I'll regard mixing in the gas phase to be so thorough that the gas remains uniform in composition (all the way down to the catalyst surface), and I'll suppose that the distribution of the various adsorbates on the catalyst surface remains uniform as well.

I have only painted a rough picture of the mechanism whereby A and B come together to form C. The details are important, and chemists work hard to try to determine what they are. There are very fine details, such as the way in which the catalyst actually makes and breaks chemical bonds, but those are not the details I mean.

When a chemist speaks of a *mechanism* for the reaction $A + B \rightarrow C$, he or she usually means a scenario of the kind depicted in (3.2), (3.3) or (3.4). Here S denotes a vacant active site on the catalyst surface. (Think of sites as atom-sized protuberances.) In (3.2), for example, the first two lines represent (reversible)

$$
\begin{aligned}
A + S &\rightleftarrows AS \\
B + S &\rightleftarrows BS \\
AS + BS &\rightarrow C + 2S
\end{aligned}
$$

(3.2)

$$
\begin{aligned}
A + S &\rightleftarrows AS \\
B + 2S &\rightleftarrows BS_2 \\
AS + BS_2 &\rightarrow C + 3S
\end{aligned}
$$

(3.3)

$$
\begin{aligned}
A + S &\rightleftarrows AS \\
B + S &\rightleftarrows BS \\
A + BS &\rightarrow C + S
\end{aligned}
$$

(3.4)

binding of A and B from the gas phase onto single vacant sites. The third line represents a surface reaction whereby the product C is formed and immediately enters the gas phase, leaving behind two vacant sites. In (3.3), B binds from the gas phase onto a pair of vacant sites rather than onto one. In both (3.2) and (3.3), the product C results from a reaction between two surface-bound species. In (3.4), on the other hand, C results from a gas phase attack of A on surface-bound B. (In this case, the adsorption of A on the surface does nothing but block active sites that might otherwise be occupied by B.)

There are many different mechanistic scenarios that can be written. Which (if any) fit the experimental facts will depend upon the catalyst and upon the reactants under study. Thus, for example, one might want to determine, in light of experimental evidence, a mechanism for the formation of carbon dioxide from carbon monoxide and oxygen in the presence of platinum.

Elementary steps of the kind shown in (3.2) – (3.3) are molecular events, and so they cannot be observed directly (at least not today). On the other hand, different mechanisms have different consequences for reactor behavior in the large, and so one can try to extract mechanistic information from whatever coarse-grained measurements are practical.

Mediating between mechanistic conjecture and experimental observations are the differential equations peculiar to each mechanism. The equations shown in (3.5), for example, are those we might write for mechanism (3.2). In (3.5) c_A^f and

$$
\begin{aligned}
c_A' &= (1/\theta)(c_A^f - c_A) - k_1 c_A c_S + k_2 c_{AS} \\
c_B' &= (1/\theta)(c_B^f - c_B) - k_3 c_B c_S + k_4 c_{BS} \\
c_C' &= -(1/\theta)c_C + k_5 c_{AS} c_{BS} \\
c_S' &= -k_1 c_A c_S + k_2 c_{AS} - k_3 c_B c_S + k_4 c_{BS} + 2k_5 c_{AS} c_{BS} \\
c_{AS}' &= k_1 c_A c_S - k_2 c_{AS} - k_5 c_{AS} c_{BS} \\
c_{BS}' &= k_3 c_B c_S - k_4 c_{BS} - k_5 c_{AS} c_{BS}
\end{aligned}
$$

(3.5)

c_B^f are the molar concentrations of A and B in the feed, c_A, c_B, c_C are the molar concentrations of A, B and C in the vessel (and in the effluent stream), c_{AS}, c_{BS} and c_S are the concentrations of AS, of BS and of vacant sites on the catalyst surface, k_1 through k_5 are rate constants for the five reactions in (3.2), and θ is the residence time for the reactor. Once again, the residence time is the volume of the mixture filling the vessel divided by the flow rate of the feed and effluent streams. (I am assuming here that the two streams are made up mostly of inert carrier gas so that their volumetric flow rates are identical.)

The dynamical equations that correspond to mechanisms (3.3) and (3.4) are similar in form but differ in detail. The rate of the step in (3.3) whereby B binds from the gas to the catalyst would, for example, be $k_3 c_B c_S^2$ rather than $k_3 c_B c_S$. The rate of the step that produces C in (3.4) would be $k_5 c_A c_{BS}$ rather than $k_5 c_{AS} c_{BS}$.

It is the difference in the dynamical equations for the various candidate mechanisms that provides a basis for favoring one over another. If gross experimental observations are incompatible with the equations induced by a particular candidate mechanism, then that mechanism can only be regarded as suspect.

But what "gross experimental observations" are practical?

Of the six dependent variables appearing in (3.5) (and in the analogous equations for the other mechanisms), only the gas phase concentrations c_A, c_B and c_C can be accessed readily. Concentrations of species adsorbed on the catalyst surface are much more difficult to determine with any degree of precision. Moreover, static measurements are much easier to make than transient ones (especially in fast processes), and so distinctions that can be drawn on the basis of steady state data are to be preferred.

Of the parameters appearing in the system (3.5) (and in its analogs for the other mechanisms) only the residence time and the feed concentrations c_A^f and c_B^f are accessible directly. (We cannot expect to know rate constants for reactions whose very existence is in question.) It is apparent, then, that distinctions between candidate mechanisms can be made only on the basis questions of the following kind: For which mechanisms are there rate constant values such that the corresponding differential equations are consistent with whatever concentration measurements are available?

My purpose here is to indicate how reaction network theory provides means to assess the viability of candidate mechanisms, even on the basis surprisingly limited experimental information. In particular, I am going to discuss ways in which mechanism discrimination can be based on observations of multiple rest points.

Let me begin with a statement that might seem surprising in view of comments made at the close of the last section: Multiple rest points (corresponding to different initial conditions) are very commonly observed in isothermal CFSTRs involving heterogeneous catalysis–that is, in reactors of the kind we are considering now.

At the end of Section 2, I said that multiple rest points are observed only rarely in isothermal homogeneous CFSTRs, so I should try to explain why things are different for catalytic CFSTRs. My claim there was that, for a homogeneous CFSTR in which the chemistry is randomly selected, multiple rest points would most likely be precluded by Theorems 2.1 and 2.2. For two reasons, this assertion is not entirely pertinent to catalytic CFSTRs: First, arguments underlying the two theorems are based on the supposition that every species in the reactor is also in the effluent stream. This is not true for catalytic CFSTRs because only the gas phase species leave the reactor. Second, reaction networks that arise in catalytic CFSTRs are far from random. Rather, they are heavily skewed toward mechanisms of the kind displayed in (3.2) – (3.4).

The theory in Section 2 was tailored very specifically to homogeneous CFSTRs. Now I want to tell you about another theory, called *deficiency one theory* [6,7], that was not developed especially for catalytic CFSTRs but which turns out to be extremely well suited to them. (On the other hand, deficiency one theory is not so good for homogeneous CFSTRs. The difference in effectiveness of the theory for the two kinds of reactors is explained in [9].)

I will not try to indicate, as I did for the SCL Graph theory, how to work with deficiency one theory. Here I can only hope to tell you something about what the theory can do, especially in connection with mechanism discrimination problems.

Although deficiency one theory is not as flashy as the SCL Graph theory, it gives much more detailed information. In rough terms, reaction networks are classified according to an easily calculated non-negative integer index called the *deficiency* of the network. If the deficiency is zero, then the corresponding mass action equations can admit only very dull, stable behavior. In particular, multiple positive rest points are impossible. (This result has its roots in work of Horn and Jackson [8].) If the deficiency turns out to be one, which is the case for catalytic CFSTRs more often than not, then the theory provides, among other things, a systematic procedure for determining – either affirmatively or negatively – whether multiple positive rest points are possible. When multiple rest points are admitted, deficiency one theory gives detailed information about relationships that rest points bear to each other.

Now I am going to suppose that two rest points (corresponding to different initial conditions) have been observed in the reactor under consideration, and I want to ask how these observations can help in mechanism discrimination. In particular, I will be interested in the extent to which the experimental facts–taken with deficiency one theory–can shed light on the viability of mechanisms (3.2) - (3.4).

In order to make the "experimental facts" concrete, I'll assume that the residence time, θ , is 1.0 and that the concentrations of A and B in the feed are $c_A^f = 1.0$ and $c_B^f = 0.8$. Furthermore, I am going to suppose that the composition of effluent gas has been determined in each of the steady states and that the results are those shown in Table 1. Moreover, I 'll suppose that a sensor indicates the presence of more A on the catalytic surface in one steady state than in the other. This too is shown in the table.[#]

****

Table 1

Gas Phase Data

Steady State No. 1	Steady State No. 2
$c_A^* = 0.6$	$c_A^{**} = 0.3$
$c_B^* = 0.4$	$c_B^{**} = 0.1$
$c_C^* = 0.4$	$c_C^{**} = 0.7$

Surface Data

More A on surface in Steady State No. 1 than in Steady State No. 2.

****

[#] The table is hypothetical, but it captures the essential features of real experiments. Measurements of the kind described were made, for example, by M. - H. Yue [15] in studies of the formation of ethane from ethylene and hydrogen on rhodium. Electrochemical techniques developed by my colleague, Howard Saltsburg, permit the assessment of relative amounts of hydrogen on the rhodium surface in the two steady states.

We might hope that distinctions between (3.2) - (3.4) could be drawn solely on the basis of the capacity of the corresponding differential equations to admit multiple rest points.

Consider, for example, mechanism (3.2). Adorned with values specified for the residence time and feed concentrations, the dynamical equations in (3.5) yield the steady state equations shown in (3.6). In this case, we would ask if there are rate

(3.6)
$$0 = (1/1.0)(1.0 - c_A) - k_1 c_A c_S + k_2 c_{AS}$$
$$0 = (1/1.0)(0.8 - c_B) - k_3 c_B c_S + k_4 c_{BS}$$
$$0 = -(1/1.0)c_C + k_5 c_{AS} c_{BS}$$
$$0 = -k_1 c_A c_S + k_2 c_{AS} - k_3 c_B c_S + k_4 c_{BS} + 2k_5 c_{AS} c_{BS}$$
$$0 = k_1 c_A c_S - k_2 c_{AS} - k_5 c_{AS} c_{BS}$$
$$0 = k_3 c_B c_S - k_4 c_{BS} - k_5 c_{AS} c_{BS}$$

constant values such that (3.6) admits two different solutions consistent with the conservation of catalyst sites. By this I mean that the total concentration of catalyst sites, occupied or not, should be the same in both steady states. In more precise terms, the two steady states should satisfy the condition

(3.7) $$c_S^* + c_{AS}^* + c_{BS}^* = c_S^{**} + c_{AS}^{**} + c_{BS}^{**}.$$

If no such rate constants exist, then mechanism (3.2) can only be regarded as suspect.

In a similar way, we can study the steady state equations for mechanisms (3.3) and (3.4) (along with the appropriate site conservation condition) to determine if they have the capacity to admit two positive rest points.

Deficiency one theory, however, indicates that mechanisms (3.2) - (3.4) *all* have the capacity for multiple rest points in an isothermal CFSTR context. Thus, *the simple qualitative capacity for multiple rest points cannot provide a basis for preferring one mechanism to another.* (In fact, deficiency one theory indicates that virtually all of the classical catalytic mechanisms have this same capacity. Remember what I said before: Observations of multiple rest points in isothermal catalytic CFSTRs are not at all unusual.)

But to say that a mechanism has the capacity to generate two steady states is not to say that it has the capacity to generate *any* two steady states. There is a sharper question we can ask, one related to the seemingly meager quantitative information that Table 1 carries: *For which of the mechanisms (3.2) - (3.4) are there rate constant values such that the corresponding steady state equations admit two solutions consistent with site conservation and with the entries in Table 1?* It will turn out that this question is surprisingly incisive.

And deficiency one theory gives the answer:

In the case of mechanism (3.2), for example, the theory indicates that there is no set of (positive) rate constant values for which (3.6) admits two solutions that are consistent with site conservation and with the information in Table 1. In particular, no matter how rate constants are assigned, there are no values of the (unmeasured) surface concentrations c_{AS}, c_{BS} and c_S in the two steady states which are simultaneously compatible with site conservation, with Table 1, and with (3.6).

The results for mechanism (3.4) are even more striking. In this case, the corresponding steady state equations cannot even admit two solutions that are consistent with site conservation and with the *gas phase* data in Table 1. In other words, *even if the surface information in Table 1 were unavailable, mechanism (3.4) would still be suspect.*

The situation for mechanism (3.3) is very different. Deficiency one theory indicates that there are values for the (unknown) rate constants and two surface compositions, $(c^*_{AS}, c^*_{BS2}, c^*_S)$ and $(c^{**}_{AS}, c^{**}_{BS2}, c^{**}_S)$, such that

(i) $c^*_{AS} > c^{**}_{AS}$

(ii) Site conservation is respected. That is,

$$c^*_{AS} + 2c^*_{BS2} + c^*_S = c^{**}_{AS} + 2c^{**}_{BS2} + c^{**}_S.$$

(iii) With (c^*_A, c^*_B, c^*_C) and $(c^{**}_A, c^{**}_B, c^{**}_C)$ as in Table 1, both

$$(c^*_A, c^*_B, c^*_C, c^*_{AS}, c^*_{BS2}, c^*_S) \text{ and } (c^{**}_A, c^{**}_B, c^{**}_C, c^{**}_{AS}, c^{**}_{BS2}, c^{**}_S)$$

are solutions of the steady state equations that correspond mechanism (3.3).

Moreover, the theory is constructive. That is, it actually gives sample values of the rate constants and of the surface concentrations in the two steady states such that (i) - (iii) are satisfied simultaneously.

The example is, I think, surprising. At first glance, Table 1 does not seem to carry much information, but it turns out to be enough to discriminate between the very similar mechanisms (3.2) - (3.4). If there is a message here, it is that even fragmentary macroscopic data may carry powerful clues about the underlying microscopic machinery.

The story I told was a simple one, but it is a reasonable reflection of the situation actually encountered by M.- H. Yue in a study of ethylene hydrogenation on rhodium [15]. In fact, deficiency one theory proved to be even more incisive there than in the example: Although there is an extensive (and contentious) literature on ethylene hydrogenation, none of the standard mechanistic proposals could account for the pairs of steady states Dr. Yue observed (even when allowances were made for gas phase mass transfer resistance and for small differences between the volumetric flow rates in the feed and effluent streams). Consistency with the data could be realized only after the construction of speculative models involving formation of ethylidine islands on the catalyst surface.

4. Traveling composition waves on isothermal catalyst surfaces. The picture of heterogeneous catalysis I painted in the last section was a very simple one, but it reflects the level of description at which modeling often takes place. Now I will consider phenomena that are more complicated, involving coupled reaction and diffusion on the catalyst surface. In particular, I want to discuss work with David Terman on possibility of traveling composition waves.

We are motivated in part by experiments of Cox, Ertl and Imbihl [4]. During the oxidation of carbon monoxide on platinum, they observed crystallographic changes in the catalyst which advanced along the surface in a wave-like fashion. These transformations are thought to be triggered by the presence of carbon monoxide, and so the observed waves probably indicate wave-like behavior in the surface concentrations of carbon monoxide and oxygen as well. On the other hand, changes in the crystal structure of platinum are almost certainly accompanied by a shift in catalytic activity. The picture that emerges is a very complex one, involving an interplay of adsorption from the gas phase, reactions on the catalyst surface, diffusion along the surface, and transformations of the catalyst itself. Cox, Ertl and Imbihl constructed a reaction-diffusion model that incorporated these effects, and, in numerical simulations, wave-like behavior was indicated.

But there remain questions of cause and effect. Although changes in the platinum crystal structure appear to be a real component of carbon monoxide oxidation, this is not to say that crystallographic shifts are intrinsic to catalysis in general. Should we, in other systems, regard the presence of composition waves as a signal that some kind of transfiguration is at work in the catalyst? Or can wave-like behavior emerge as a natural consequence of "normal" catalysis, divorced from ancillary effects?

Terman and I posed for ourselves these questions: *Can traveling composition waves derive solely from catalytic mechanisms as simple as those shown in (3.2) - (3.4) (taken only with diffusion along the catalyst surface)? If so, which mechanisms are consistent with the existence of composition waves? In particular, is the capacity for traveling waves in some sense a generic property, shared by a wide spectrum of mechanisms?*

Our interest is in fundamental occurrences on the catalyst surface itself, uncomplicated by dynamics in the gas phase. With this in mind, we depart from the CFSTR configuration studied in the last section and consider instead a catalytic surface immersed in a spatially homogeneous gaseous environment of time-invariant composition. This is very pristine catalysis.

In this context, we consider the large family of mechanisms shown in (4.1). Here

(4.1)
$$A + mS \rightleftarrows kX$$
$$B + nS \rightleftarrows jY$$
$$pX + qY \rightarrow (p(m/k) + q(n/j))S + rC$$

A, B and C are gas-phase species. The first line represents binding of A from the gas phase to m vacant catalyst sites and the subsequent formation of k copies of the surface species X. (Each X occupies m/k sites.) The situation in the second line is similar: A molecule of B binds to n sites, and j copies of the surface species Y are formed. In the last line, p copies of X react with q copies of Y to form r molecules of the product C, which enter the gas phase, leaving behind $p(m/k) + q(n/j)$ sites.

Note that the mechanisms (3.2) - (3.3) studied in the last section are typical members of the family (4.1). So are (4.2) and (4.3), which are mechanisms one might consider as candidates for the oxidation of carbon monoxide. In (4.2) and (4.3) the gaseous reactants A and B are carbon monoxide and oxygen, and the product C is carbon dioxide.

$$CO + S \rightleftarrows CO \cdot S$$
(4.2) $$O_2 + 2S \rightleftarrows 2O \cdot S$$
$$CO \cdot S + O \cdot S \rightarrow CO_2 + 2S$$

$$CO + S \rightleftarrows CO \cdot S$$
(4.3) $$O_2 + S \rightleftarrows O_2 \cdot S$$
$$2CO \cdot S + O_2 \cdot S \rightarrow 2CO_2 + 3S$$

The (constant) concentrations of A and B in the gas phase will be denoted by c_A^* and c_B^*, and the five rate constants for the reactions in (4.1) will be denoted (from top to bottom) by $\alpha, \beta, \gamma, \varepsilon$, and ξ. The concentrations of vacant sites and of the surface species X and Y are the ones we want to study. Were these concentrations to remain spatially uniform, they would be governed by the system of ordinary differential equations shown in shown (4.4).

(4.4a) $\quad c_X' = k\alpha c_A^*(c_S)^m - k\beta(c_X)^k - p\xi(c_X)^p(c_Y)^q$

(4.4b) $\quad c_Y' = j\gamma c_B^*(c_S)^n - j\varepsilon(c_Y)^j - q\xi(c_X)^p(c_Y)^q$

(4.4c) $\quad c_S' = -m\alpha c_A^*(c_S)^m + m\beta(c_X)^k$
$$- n\gamma c_B^*(c_S)^n + n\varepsilon(c_Y)^j$$
$$+ (p(m/k) + q(n/j))\xi(c_X)^p(c_Y)^q$$

Note that
$$(m/k)c_X' + (n/j)c_Y' + c_S' = 0.$$

Along solutions of (4.4), therefore, the value of

$$(m/k)c_X + (n/j)c_Y + c_S$$

is invariant. In fact, this sum is just the concentration of catalyst sites, occupied and unoccupied. (Recall that X and Y occupy (m/k) and (n/j) sites, respectively.) With this in mind, we restrict our attention to solutions of (4.4) consistent with the equation

(4.5) $$(m/k)c_X + (n/j)c_Y + c_S = K,$$

where the (positive) number K is to be interpreted as the site concentration for the catalyst surface under study.

In connection with our study of traveling waves, we will want to know which members of the family (4.1) have the capacity to give multiple positive rest points. That is, we will want to know for which positive values of the stoichiometric coefficients m, k, n, j, p and q are there positive values of the parameters $\alpha c_A^*, \beta, \gamma c_B^*, \varepsilon, \xi$, and K such that (4.4) has two or more positive rest points compatible with (4.5). (Remember that the setting here is different from the CFSTR setting studied in Section 3.)

The answer is again given by deficiency one theory [14]:

THEOREM 4.1. *If $m \neq n$, then (regardless of the positive values that the stoichiometric coefficients k, j, p and q take) there are positive values for the parameters $\alpha c_A^*, \beta, \gamma c_B^*, \varepsilon, \xi$, and K such that (4.4) admits two or more positive rest points compatible with (4.5). If $m = n$, then parameter values of this kind exist if and only if*

$$m \neq p(m/k) + q(n/j)$$

and either $q > j$ or $p > k$ (or both).

For some special cases of (4.1), all with $m = k$ and $n = j$, the possibility of multiple rest points was already studied by Bykov, Chumakov, Elokhin and Yablonskii [2].

Theorem 4.1 indicates that, for the special physical situation under consideration here, the capacity for multiple rest points is robust within the mechanistic family (4.1). What we want to argue is that, within the same mechanistic family, the capacity for stable traveling waves along the catalyst surface is about as robust as the capacity for multiple rest points.

In preparation for formulation of reaction-diffusion equations, note that, from (4.5), c_S can be written in terms of c_X and c_Y, and the result can be inserted in (4.4a) and (4.4b). Then (4.4a) and (4.4b) can be viewed as a self-contained pair of ordinary differential equations that govern c_X and c_Y when those concentrations remain spatially uniform over the catalyst surface.

Our interest will be in distributions of X and Y that are not necessarily uniform. If we assume–as Cox, Ertl and Imbihl did–that diffusive transport along the catalyst surface is of the simple Fickian kind[#] , then the (one-dimensional) reaction-diffusion

---

[#] For the purposes of this exploratory study, it is natural to follow Cox, Ertl and Imbihl in considering simple Fickian diffusion. Indeed, the model is plausible when the catalyst surface is not crowded with adsorbed species. When the surface is crowded, however, it seems unreasonable to expect the local surface diffusion of a species to depend only on the local concentration gradient of that particular species, independent of the availability of nearby vacant sites. In this case, models of the kind considered in [1] would seem more apt.

analogs of (4.4a) and (4.4b) are the equations shown in (4.6).

$$(c_X)_t = D_X(c_X)_{zz} + k\alpha c_A^*(K - (m/k)c_X - (n/j)c_Y)^m$$
$$- k\beta(c_X)^k - p\xi(c_X)^p(c_Y)^q$$

(4.6)

$$(c_Y)_t = D_Y(c_Y)_{xx} + j\gamma c_B^*(K - (m/k)c_X - (n/j)c_Y)^n$$
$$- j\varepsilon(c_Y)^j - q\xi(c_X)^p(c_Y)^q$$

Now I am going to consider some fixed member of the mechanistic family (4.1) that is guaranteed by Theorem 4.1 to admit multiple positive rest points in (4.4) - (4.5). I shall suppose that the parameters $\alpha c_A^*, \beta, \gamma c_B^*, \varepsilon, \xi$, and K take values such that multiple positive rest points are, in fact, admitted and such that all the positive rest points are nondegenerate. The claim I want to make is this: These parameter values can be supplemented with values of the diffusivities D_X and D_Y in such a way that the corresponding reaction-diffusion equations (4.6) admit a stable traveling wave.

This is summarized in a theorem due to David Terman [14]:

THEOREM 4.2. *Consider any member of the mechanistic family (4.1) for which there are values of $\alpha c_A^*, \beta, \gamma c_B^*, \varepsilon, \xi$, and K that give multiple positive rest points in (4.4) - (4.5). If, for such a set of parameter values, all positive rest points of (4.4) - (4.5) are nondegenerate, then there are values of the diffusivities D_X and D_Y for which the corresponding reaction-diffusion equations (4.6) admit a stable traveling wave. In fact, it suffices to take $D_X = D_Y > 0$.*

As I indicated in the Introduction, the proof rests on Conley Index methods [3]. The waves Terman showed to exist are of the kind that travel from a region at $z = -\infty$ to a region at $z = +\infty$, where the two regions are characterized by compositions corresponding to stable rest points of (4.4) - (4.5).

Terman showed even more: For some members of the family (4.1), studies of Bykov, Chumakov, Elokhin and Yablonskii [2] indicate the existence parameter values that give rise to more than two stable rest points of (4.4) - (4.5). In such instances, Terman proved that the stable rest points can be sequenced in such a way that there is a stable traveling wave connecting consecutive members of the sequence, and, moreover, such that wave speeds are nonincreasing along the sequence.

Taken together, Theorems 4.1 and 4.2 suggest that traveling waves are probably unexceptional (if not commonplace) in "normal" catalysis, the absence of crystallographic transitions notwithstanding: Theorem 4.1 indicates that, for the mechanistic family and the physical setting under consideration here, the capacity for multiple positive rest points is rather robust. For any mechanism that does have such a capacity, Theorem 4.2 indicates that, for almost all parameter values for which multiple positive rest points are admitted by the ordinary differential equations, there are diffusivities for which stable traveling waves are admitted by the corresponding reaction-diffusion equations.

REFERENCES

[1] BYKOV, V., A. GORBAN, L. KAMENSCHCHIKOV, AND G.S. YABLONSKII, *Inhomogeneous stationary states in the reaction of carbon monoxide on platinum*, Kinetics and Catalysis, 24 (1983) pp. 520–524.

[2] BYKOV, V. I., G. A. CHUMAKOV, V. I. ELOKHIN, G. S. YABLONSKII, *Dynamic properties of a heterogeneous catalytic reaction with several steady states*, Reaction Kinetics and Catalysis Letters, 4 (1976), pp. 397–403.

[3] CONLEY, CHARLES, *Isolated Invariant Sets and the Morse Index*, CBMS Regional Conference Series in Mathematics, No. 38, American Mathematics Society (1978).

[4] COX, M.P., G. ERTL, AND R. IMBIHL, *Spatial self-organization of surface structure during an oscillating catalytic reaction*, Physical Review Letters, 54 (1985), pp. 1725–1728.

[5] FEINBERG, M., *Chemical Oscillations, Multiple Equilibria, and Reaction Network Structure*, in Dynamics and Modelling of Reactive Systems, eds. Warren Stewart, W. Harmon Ray, and Charles Conley, Academic Press, New York, 1980, pp. 59–130.

[6] FEINBERG, M., *Chemical reaction network structure and the stability of complex isothermal reactors: I. The deficiency zero and deficiency one theorems*, Chemical Engineering Science, 42 (1987), pp. 2229–2268.

[7] FEINBERG, M.,, *Chemical reaction network structure and the stability of complex isothermal reactors: II. Multiple steady states for networks of deficiency one*, Chemical Engineering Science, 43 (1988), pp. 1–25.

[8] HORN, F.J.M. AND ROY JACKSON, *General mass action kinetics*, Archive for Rational Mechanics and Analysis, 47 (1972), pp. 81–116.

[9] LEIB, T.M., D. RUMSCHITZKI AND M. FEINBERG, *Multiple steady states in complex isothermal CFSTRs: I. General Considerations*, Chemical Engineering Science, 43 (1988), pp. 321–328.

[10] RUMSCHITZKI, D. AND M. FEINBERG, *Multiple steady states in complex isothermal CFSTRs: II. Homogeneous reactors*, Chemical Engineering Science, 43 (1988), pp. 329–337.

[11] SCHLOSSER, P., *A Graphical Determination of the Possibility of Multiple Steady States in Complex Isothermal CFSTRs*, PhD Thesis, Department of Chemical Engineering, University of Rochester (1988).

[12] SCHLOSSER, P. AND M. FEINBERG, *A graphical determination of the possibility of multiple steady states in complex isothermal CFSTRs*, pp. 102-115 in Complex Chemical Reaction Systems, eds. J. Warnatz and W. Jäger, Springer-Verlag, Berlin-Heidelberg-New York, 1988.

[13] SCHLOSSER, P. AND M. FEINBERG, *Multiple steady states in very complex isothermal CFSTRs*, in preparation for Chemical Engineering Science.

[14] TERMAN, D. AND M. FEINBERG, *Traveling waves on isothermal catalyst surfaces*, in preparation for Archive for Rational Mechanics and Analysis.

[15] YUE, M.-H., *Isothermal Multiple Steady States in Ethylene Hydrogenation over Rhodium: Mechanistic Screening and Experimental Studies*, PhD Thesis, Department of Chemical Engineering, University of Rochester (1989).

GENERICITY, BIFURCATION AND SYMMETRY

MARTIN GOLUBITSKY*

In these lectures I would like to discuss how the existence of symmetries alters the type of bifurcation behavior that one expects to observe. In the first lecture I will concentrate on the structure and *dynamics* of steady-state bifurcation from equilibria. It is here that the influence of symmetries on linearized equations will be discussed and some facts from elementary representation theory introduced. The second lecture will be devoted to effects of symmetry on period-doubling in maps with a short description of an application to large arrays of Josephson junctions. In the final lecture I will describe how certain standard choices of boundary conditions (particularly Neumann) can be thought of as symmetry constraints and how this fact alters notions of genericity. It accord with the style that has developed in the lectures at this workshop, the lectures are of different length.

Much of the background material for these lectures may be found in [GSS]. The descriptions of the more advanced topics will be brief as the results concerning these topics have or will appear elsewhere.

Lecture 1: *Bifurcation From Equilibria*

Consider the system of ODE

$$(1.1) \qquad \frac{dx}{dt} = f(x, \lambda) \qquad x \in \mathbf{R}^n , \ \lambda \in \mathbf{R}$$

with an equilibrium at (x_0, λ_0), that is,

$$f(x_0, \lambda_0) = 0.$$

Let $A \equiv (df)_{x_0, \lambda_0}$ be the $n \times n$ Jacobian matrix obtained by differentiation with respect to x. Then (1.1) becomes:

$$\frac{dx}{dt} = A(x - x_0) + \cdots$$

Recall that if A is hyperbolic (that is, all eigenvalues have nonzero real part), then all the dynamics of (1.1) are determined by A near x_0.

*Department of Mathematics, University of Houston, Houston, TX 77204–3476

DEFINITION 1.1. (1.1) has a *bifurcation point* at (x_0, λ_0) if some eigenvalue of A lies on the imaginary axis.

Without loss of generality we may assume $\lambda_0 = 0$.

Basic Question: What are the typical transitions in the dynamics of (1.1)?

It is well known that generically the typical transitions are controlled by the transitions from hyperbolicity in the matrix A as λ is varied. Indeed, there are two possibilities; A has a

(a) simple zero eigenvalue   *Steady-State Bifurcation*

(b) a pair of simple, complex-conjugate, purely imaginary eigenvalues.   *Hopf bifurcation*

Nonlinear theory then implies that in case (a) the dynamics can be reduced (using center manifolds) to one dimension and the expected transition is a limit-point or saddle-node bifurcation with a transition from 0 to 2 equilibria. In case (b) the dynamics can be reduced to two dimensions and one expects a single branch of periodic solutions to emanate from this bifurcation. See [GH1].

We now consider how both the linear and the nonlinear transitions change when (1.1) has a nontrivial group of symmetries.

SYMMETRY

Let $\Gamma \subset O(n)$ be a Lie group of orthogonal matrices.

DEFINITION 1.2. (1.1) has *symmetry* Γ if

$$(1.2) \qquad\qquad f(\gamma x, \lambda) = \gamma f(x, \lambda) \qquad \text{for all} \quad \gamma \in \Gamma.$$

There are two immediate consequences of the commutativity relation (1.2):

(a) The equilibrium x_0 has *symmetry*. Define the *isotropy subgroup* of Γ at x_0 to be

$$\Sigma_{x_0} \equiv \{\gamma \in \Gamma : \gamma x_0 = x_0\}$$

In our discussion we will assume that the equilibrium x_0 is a fully symmetric, i.e. $\Sigma_{x_0} = \Gamma$. Then, without loss of generality, we may assume $x_0 = 0$.

(b) The chain rule implies $(df)_{\gamma x, \lambda}\gamma = \gamma(df)_{x, \lambda}$ and hence

$$A\gamma = \gamma A.$$

that is, the matrix A *commutes* with Γ. Thus to understand the dynamics of (1.1) we must first understand the form of matrices that commute with Γ. This topic in representation theory has been well studied and we briefly review the relevant theory.

ELEMENTARY REPRESENTATION THEORY

DEFINITIONS 1.3. Let W be a subspace of $V \cong \mathbf{R}^n$.

(a) $W \subset V$ is Γ-*invariant* if $\gamma(W) = W$ for all $\gamma \in \Gamma$

(b) $W \subset V$ if Γ-*irreducible* if the only Γ-invariant subspaces of W are $\{0\}$ and W.

It is well known that any representation may be decomposed into a direct sum of irreducible representations; the simplicity of the proof of this decomposition is, however, not always appreciated.

THEOREM 1.4 (*The Decomposition Theorem*). *There exist* Γ-*irreducible subspaces* $V_1, \ldots, V_s$ *such that*
$$V = V_1 \oplus \cdots \oplus V_s .$$

Proof. Since $\Gamma \subset O(n)$, the standard inner product is Γ-invariant; that is, $(\gamma v, \gamma w) = (v, w)$ for all $\gamma \in \Gamma$.

Now suppose V has proper Γ-invariant subspace W. Then define

$$W^\perp = \{v \in V : (v, W) = 0\}$$

and observe that $W^\perp$ is Γ-invariant. Since

$$V = W \oplus W^\perp.$$

the theorem is proved by induction on the dimension of V. $\qquad\square$

We begin our discussion of commuting matrices by first considering commuting matrices for an irreducible representation U.

THEOREM 1.5. *The space of matrices commuting with an irreducible representation is isomorphic to* $\mathbf{R}, \mathbf{C}$ *or* $\mathbf{H}$ (*where* $\mathbf{H}$ *denotes the quaternions*).

Proof. Observe that the vector space

$$\mathcal{D} = \{\text{matrices on } U \text{ commuting with } \Gamma\}$$

is an algebra over $\mathbf{R}$; that is, we can add, multiply and scalar multiply commuting matrices. In addition, $\mathcal{D}$ is a division algebra, that is, every nonzero matrix B in $\mathcal{D}$ is invertible. To verify this point note that for $B \in \mathcal{D}$

$$\ker B \text{ is } \Gamma\text{-invariant}$$

and irreducibility implies

$$\ker B = U \text{ or } \ker B = \{0\}$$

Hence, either $B = 0$ or B is invertible and $B^{-1} \in \mathcal{D}$. The classical Wedderburn Theorem implies that $\mathcal{D}$ is isomorphic either to $\mathbf{R}, \mathbf{C}$ or $\mathbf{H}$.

DEFINITION 1.6. U is *absolutely* irreducible if the only matrices commuting with Γ are multiples of the identity ($\mathcal{D} \cong \mathbf{R}$) and *nonabsolutely* irreducible otherwise.

Examples. (a) $SO(2)$ acts nonabsolutely irreducibly on $\mathbf{R}^2$.

(b) $O(2)$ acts absolutely irreducibly on $\mathbf{R}^2$.

THEOREM 1.7. *Generically, in one-parameter bifurcation, steady-state bifurcation satisfies:*

(a) *the algebraic multiplicity of the zero eigenvalue equals the geometric multiplicity, and*

(b) Γ *acts absolutely irreducibly on* ker A.

Sketch of Proof.

(I) At a bifurcation point, do a center manifold reduction (which can be done preserving symmetries - Ruelle [R]). Thus, we can assume all eigenvalues of A are on the imaginary axis.

(II) Suppose A has a zero eigenvalue.

Choose a Γ-irreducible subspace $U \subset$ ker A, and define

$$M : V \to V \quad \text{by} \quad \begin{cases} 0 & \text{on} & U \\ I & \text{on} & U^\perp \end{cases}$$

Perturb (1.1) to:
$$\frac{dx}{dt} = f(x, \lambda) + \varepsilon M x \equiv f_\varepsilon(x, \lambda).$$

For nonzero ε, $A_\varepsilon \equiv (df_\varepsilon)_{0,0} = A + \varepsilon M$ satisfies:

(a) Geometric multiplicity of eigenvalue zero

$\qquad\qquad$ = algebraic multiplicity of eigenvalue zero.

(b) Γ acts irreducibly on ker $A_\varepsilon = U$.

Now reduce the bifurcation problem to U (by center manifold).

(III) If dim $U = 1$, then Γ acts absolutely irreducibly. So assume dim $U \geq 2$. We claim that $f(0, \lambda) = 0$, that is, $x = 0$ is a 'trivial' equilibrium.

DEFINITION 1.8. Let $\Sigma \subset \Gamma$ be a subgroup. Define the *fixed-point subspace* of Σ to be:
$$\text{Fix}(\Sigma) \equiv \{v \in V : \sigma v = v \text{ for all } \sigma \in \Sigma\}$$

LEMMA 1.9. $f : \text{Fix}(\Sigma) \times \mathbf{R} \to \text{Fix}(\Sigma)$.

Proof. $f(v, \lambda) = f(\sigma v, \lambda) = \sigma f(v, \lambda)$ for all $\sigma \in \Sigma$.

$\qquad\qquad\qquad\qquad\qquad\qquad\qquad\qquad\qquad\qquad\qquad$ □

To prove the claim observe that $\text{Fix}(\Gamma) = \{0\}$ since Γ acts irreducibly and nontrivially. Thus, $f(0, \lambda) = 0$.

Define $A_\lambda = (df)_{0,\lambda}$ and observe that A_λ commutes with Γ. Hence, A_λ is in $\mathcal{D}$ and corresponds to a curve $d(\lambda) \in \mathbf{R}, \mathbf{C}$ or $\mathbf{H}$ with $d(0) = 0$.

(IV)   Suppose that Γ acts nonabsolutely irreducibly on U. Then the curve $d(\lambda)$ is in either $\mathbf{C}$ or $\mathbf{H}$. The hyperplane $\{z \in \mathcal{D} : \text{Re}(z) = 0\}$ corresponds to matrices with purely imaginary eigenvalues. Hence the curve $d(\lambda)$ can be perturbed to $e(\lambda)$ where $e(\lambda)$ crosses $\text{Re}(z) = 0$ at $\lambda = 0$ with nonzero speed, but NOT THROUGH 0. Since the curve $e(\lambda)$ corresponds to a family of matrices B_λ, we can perturb (1.1) to:

$$\frac{dx}{dt} = f(x, \lambda) + (B_\lambda - A_\lambda)x.$$

□

The absolute irreducibility noted in Theorem 1.9 can be used to transfer the analytic problem of existence of branches of equilibria to an algebraic one, as the next theorem shows.

THEOREM 1.10 (*Equivariant Branching Lemma*). (Vanderbauwhede [V], Cicogna [C])

(a)   Let $\Gamma \subset O(n)$ be a Lie group acting absolutely irreducibly on $V \equiv \mathbf{R}^n$.

(b)   Assume that (1.1) has a bifurcation at $\lambda = 0$ and symmetry Γ.

(c)   Let $\Sigma \subset \Gamma$ be a subgroup such that

(1.3) $\dim \text{Fix}(\Sigma) = 1.$

then there is a unique branch of equilibria having symmetry Σ if

$$c'(0) \neq 0$$

where $(df)_{0,\lambda} = c(\lambda)I$ by (a) and $c(0) = 0$ by (b).

Proof. We know that $f : \text{Fix}(\Sigma) \times \mathbf{R} \to \text{Fix}(\Sigma)$. Let v_0 be a nonzero vector in $\text{Fix}(\Sigma)$, and define $g : \mathbf{R} \times \mathbf{R} \to \mathbf{R}$ by

$$g(s, \lambda)v_0 = f(sv_0, \lambda).$$

We that $g(0, \lambda) = 0$ since irreducibility implies that 0 is a 'trivial' solution. Hence $g(s, \lambda) = h(s, \lambda)s$ by Taylor's Theorem where

$$h(0,0) = c(0) = 0 \quad \text{and} \quad h_s(0,0) = c'(0) \neq 0.$$

Using the Implicity Function Theorem solve

$$h(s, \lambda) = 0 \quad \text{for} \quad \lambda = \Lambda(s).$$

□

Examples. (a) $O(2)$ acts on $\mathbf{R}^2 \cong \mathbf{C}$. Let $\Sigma = \{1, \kappa\} \cong \mathbf{Z}_2$ where $\kappa z = \bar{z}$. Then $\mathrm{Fix}(\Sigma) = \mathbf{R} \subset \mathbf{C}$ has dimension one. Hence, in circularly symmetric bifurcation problems we expect equilibria with a reflectional symmetry.

(b) The irreducible representations of $S0(3)$ are given by the spherical harmonics of order ℓ denoted by V_ℓ. (Note that $\dim V_\ell = 2\ell+1$.) The Cartan decomposition of V_ℓ is:

$$V_\ell = \mathbf{R} \oplus \mathbf{C}^\ell$$

where the action of $S0(2) \subset S0(3)$ on V_ℓ is given by:

$$\theta \cdot (x, \; z_1, \ldots, z_l) = (x, e^{i\theta} z_1, \ldots, e^{li\theta} z_l).$$

It follows that $\mathrm{Fix}(SO(2)) = \{(x, 0, \ldots, 0)\}$ has dimension one. Since solutions with $SO(2)$ symmetry have an axis of symmetry, we have proved:

COROLLARY 1.11. *In steady-state bifurcations involving spherical symmetry, generically (in the sense that eigenvalues go through zero with nonzero speed) there exist a branch of axisymmetric equilibria.*

(c) (Nontrivial *dynamics* in steady-state bifurcation) Let $\Gamma \subset O(3)$ be the 24 element group generated by:

$$\sigma(x, y, z) = (y, z, x)$$
$$\varepsilon(x, y, z) = (\varepsilon_1 x, \varepsilon_2 y, \varepsilon_3 z) \qquad \text{where} \qquad \varepsilon_j = \pm 1.$$

Bifurcation with this group action was studied by May and Leonard [ML] in the context of three competing populations and Busse and Heikes [BH] in the context of convection in a rotating layer. Later, Guckenheimer & Holmes [GH2] studied this group of symmetries abstractly. They showed that it is possible to have a structurally stable (in the world of Γ symmetry), asymptotically stable, primary branch of heteroclinic connections. We outline this construction.

Up to conjugacy the isotropy subgroups of Γ are:

$$\Sigma_2 = \{\varepsilon_1, 1, 1)\} \qquad \mathrm{Fix}(\Sigma_2) = \{(0, y, z)\}$$
$$\Sigma_3 = \{(1, \sigma, \sigma^2\} \qquad \mathrm{Fix}(\Sigma_3) = \{(x, x, x)\}$$
$$\Sigma_4 = \{(\varepsilon_1, \varepsilon_2, 1)\} \qquad \mathrm{Fix}(\Sigma_4) = \{(0, 0, z)\}.$$

Hence, generically, there exist two types of equilibria with isotropy Σ_3 and Σ_4, respectively.

We determine the dynamics associated with this Γ-equivariant bifurcation by describing explicitly the general Γ-equivariant mapping f. Write f in coordinates as $f = (X, Y, Z)$. Then

(i) $f(\sigma v, \lambda) = \sigma f(v, \lambda)$ implies :

$$Y(x, y, z, \lambda) = X(y, z, x, \lambda)$$

$$Z(x, y, z, \lambda) = X(z, x, y, \lambda).$$

((ii) $f(\varepsilon v, \lambda) = \varepsilon f(v, \lambda)$ implies

$$X(x, y, z, \lambda) \text{ is odd in } x \text{ and even in } y \text{ and } z.$$

Thus, we can write $X(x, y, z, \lambda) = a(x^2, y^2, z^2, \lambda)x$. The genericity condition is $a_\lambda(0) \neq 0$. We assume:

(H1) $a_\lambda(0) > 0$

so that the trivial solution losses stability at $\lambda = 0$. Rescale λ so that:

$$a_\lambda(0) = 1.$$

To third order f has the form:

$$\begin{pmatrix} (\lambda + \alpha x^2 + \beta y^2 + \gamma z^2)x \\ (\lambda + \gamma x^2 + \alpha y^2 + \beta z^2)y \\ (\lambda + \beta x^2 + \gamma y^2 + \alpha z^2)z \end{pmatrix} .$$

Hence, computing $f | \operatorname{Fix}(\Sigma_4) \times \mathbf{R} = 0$ yields the equation $\alpha z^2 = \lambda$. Thus, if we assume

(H2) $\alpha < 0,$

then the Σ_4 equilibrium A will exist for $\lambda > 0$ and, by exhance of stability, be stable inside the z-coordinate axis $\operatorname{Fix}(\Sigma_4)$.

Since $(df)_A$ commutes with its isotropy subgroup Σ_4. It follows that $(df)_A$ is diagonal and that, to lowest order, the two eigenvalues outside $\operatorname{Fix}(\Sigma_4)$ are:

$$(\gamma - \alpha)z^2 \qquad \text{(in the } x\text{-direction)}$$

$$(\beta - \alpha)z^2 \qquad \text{(in the } y\text{-direction)}$$

If we assume:

(H3) $\beta < \alpha < \gamma$

then A will be a sink in the flow-invariant yz-plane $\mathrm{Fix}(\Sigma_2)$ and a saddle in the xz-plane. By considering $f \mid \mathrm{Fix}(\Sigma_2) \times \mathbb{R}$ one can show that there are no equilibria in the yz-plane that lie off the coordinate axes when (H3) is valid.

Thus the unstable manifold leaving A in the xz-plane must either be unbounded or tend to the equilibrium on the x-axis. Indeed, the saddle-sink connection can be shown to exist if:

$$a \ll 0.$$

Hence the heteroclinic cycle exists.

Finally, we note that this connection is structurally stable, since the coordinate planes are always flow invariant (being fixed point subspaces) and planar saddle-sink connections are structurally stable. A calculation shows that this heteroclinic connection can be asymptotically stable.

Thus, intermittancy is an expected phenomena in symmetric systems. Further examples of complicated dynamics emanating from a steady-state bifurcation may be found in Field [F3]. See also [AGH]

We end this lecture by discussing the general form of linear commuting maps when the representation is not irreducible. This result is useful when computing the asymptotic stability of nontrivial equilibria, and when considering mode interactions in multiparameter systems. This material is included mainly for completeness and may be skipped on a first reading.

DEFINITIONS 1.12. Let Γ act on a space V.

(a) Let W_1 and W_2 be Γ-irreducible subspaces of V. Then W_1 and W_2 are Γ-isomorphic if there exists a linear map $L : W_1 \to W_2$ that commutes with Γ, that is, $L\gamma = \gamma L$ for all $\gamma \varepsilon \Gamma$.

(b) Let W be a Γ-irreducible subspace of V. The isotypic component of V corresponding to W is the sum of all Γ-irreducible subspaces of V are Γ-isomorphic to W.

Examples. (a) For each integer ℓ let $O(2)$ act on $V_\ell \cong \mathbb{C}$ by:

$$\theta \cdot z = e^{\ell i\theta} z \quad \text{and} \quad \kappa \cdot z = \overline{z}.$$

The actions for ℓ_1 and ℓ_2 are $O(2)$-isomorphic iff $\ell_1 = \pm\ell_2$.

(b) Let the permutation group S_3 act on $\mathbb{C}$ as symmetries of a equilateral triangle and on $\mathbb{R}^3$ by permuting axes. The second action has a two dimensional S_3 irreducible subspace consisting of points in $\mathbb{R}^3$ whose coordinates sum to zero. The actions of S_3 on $\mathbb{C}$ and V are isomorphic.

THEOREM 1.13. Let $U_1, \ldots, U_t$ be the distinct Γ-irreducible representations appearing in a decomposition of V guaranteed by the Decomposition Theorem. Then V is the direct sum of isotypic components:

(1.4) $$V = V_{u_1} \oplus \cdots \oplus V_{u_t}.$$

COROLLARY 1.14. *Let* $A : V \to V$ *be linear and commute with* Γ. *Then* A *can be block diagonalized by (1.4), that is,*

$$A(V_{u_j}) \subset V_{u_j} \quad for \quad j = 1, \ldots, t.$$

Lecture 2: *Period-Doubling, Symmetry and Josephson Functions*

In this lecture I want to describe how symmetry affects period-doubling bifurcations and apply these ideas to coupled systems of Josephson junctions. The theory follows closely the discussion of steady-state bifurcations given in the first lecture - but with an important difference. The period-doubling bifurcation itself introduces a reflectional symmetry.

Let $f : V \times \mathbb{R} \to V$ be a smooth Γ-equivariant mapping and let $f(\cdot, \lambda_0)$ have a fixed-point at x_0.

DEFINITION 2.1. f *has a* period-doubling *bifurcation at* (x_0, λ_0) *if* -1 *is an eigenvalue of the Jacobian matrix* $(df)_{x_0, \lambda_0}$.

We assume that x_0 is Γ-invariant and hence, without loss of generality, we may assume that $(x_0, \lambda_0) = (0, 0)$. Our discussion of genericity for steady-state bifurcation applies equally well to period-doubling bifurcation. In particular, genericity implies that the geometric multiplicity of the eigenvalue -1 equals the algebraic multiplicity and that Γ acts absolutely irreducibly on the eigenspace V_{-1} corresponding to the eigenvalue -1.

After a center manifold reduction we may assume that $V = V_{-1}$. Observe that irreducibility implies that f has a trivial fixed point, i.e. that $f(0, \lambda) = 0$. Similarly, absolute irreducibility implies that

$$(df)_{0,\lambda} = c(\lambda)I$$

where $c(0) = -1$. Indeed, genericity implies that $c'(0) \neq 0$.

The question we address is: find all branches of period two points of f in the neighborhood of the period-doubling bifurcation at $(0, 0)$. We prove the following analogue of the Equivariant Branching Lemma.

Define the group

$$\widehat{\Gamma} = \begin{cases} \Gamma & \text{if} \quad -I \in \Gamma \\ \Gamma \oplus \mathbb{Z}_2(-I) & \text{if} \quad -I \notin \Gamma. \end{cases}$$

Note that $\widehat{\Gamma}$ acts naturally on V.

THEOREM 2.2. *Let* $\Sigma \subset \widehat{\Gamma}$ *be a subgroup satisfying:*

$$\dim (\mathrm{Fix}(\Sigma)) = 1.$$

Then there exists a unique branch of period two points for f emanating from the origin.

A proof of this theorem, based on normal hyperbolicity is given in [ChG]. Here, however, we present a very simple proof using Liapunov–Schmidt reduction. This proof was derived independently by Peckham & Kevrikidis, Roberts and Vanderbauwhede (private communications). The idea of the proof is to convert the problem of finding period two points of f to one of finding zeroes of a derived mapping F. The Equivariant Branching Lemma is then used to prove the existence of branches of zeroes of F. The proof is a discrete analogue of the proof of existence of periodic solutions given by Liapunov–Schmidt reduction in Hopf bifurcation.

Proof. Observe that finding a point x such that $f(f(x)) = x$ is equivalent to finding solutions to the system of equations

$$y = f(x) \qquad \text{and} \qquad x = f(y).$$

Given this, define $F : V \times V \to V \times V$ by

$$F(x, y) = (f(x) - y, \ f(y) - x).$$

Then $F(x, y) = (0, 0)$ if and only if x and y are period two points of f.

Next use Liapunov–Schmidt reduction to solve $F = 0$. To do this, compute

$$(dF)_{0,0} = \begin{pmatrix} -I & -I \\ -I & -I \end{pmatrix}$$

and observe that

$$\widetilde{V} \equiv \ker (dF)_{0,0} = \{(x, -x) \in V \times V : x \in V\}$$

is isomorphic to V. Now use Liapunov-Schmidt reduction to find implicitly a mapping $g : \widetilde{V} \to \widetilde{V}$ whose zeroes near the origin are in one to one correspondence with the zeroes of F.

Now we consider the equivariance properties of g. Since Liapunov–Schmidt reduction can be performed in such a way as to preserve symmetries, we need only consider the equivariance of F. Note that Γ acts (via the diagonal action) on $V \times V$ and that F commutes with the action of Γ. In addition F commutes with the reflection symmetry $(x, y) \to (y, x)$. Since this symmetry acts as $-I$ on $\widetilde{V}$, it follows that the reduced mapping g commutes with the group $\widehat{\Gamma}$ acting on $\widetilde{V}$.

Finally, we note that the assumption on $\mathrm{Fix}(\Sigma)$ is precisely what is needed to apply the Equivariant Branching Lemma. $\quad\Box$

Arrays of Coupled Oscillators

We will apply this theorem to find period two points emanating from period-doubling bifurcations in the presence of S_N symmetry, where S_N is the permutation group on N letters. To motivate this discussion, we begin by considering arrays of coupled oscillators.

An array of *coupled oscillators* is a system of ODE of the form:

(2.1)
$$\dot{y}_1 = g_1(y_1, \ldots, y_N)$$
$$\ldots \qquad\qquad y_j \in \mathbf{R}^k$$
$$\dot{y}_N = g_N(y_1, \ldots, y_N).$$

These oscillators are *identical* if $g_1 = \cdots = g_N \equiv g$ and *identically coupled* if

$$g(y_1, y_2, \ldots, y_N) = g(y_1, y_{\sigma(2)}, \ldots, y_{\sigma(N)})$$

for every permutation on $N - 1$ letters σ. Observe that systems of identical, identically coupled, coupled oscillators are precisely those systems (2.1) that have S_N symmetry.

An *in-phase* solution to (2.1) is one lying in the plane

(2.3)
$$y_1 = \cdots = y_N \equiv y.$$

The fact that the plane defined by (2.3) is flow-invariant follows from the fact that the plane (2.3) is just $\text{Fix}(S_N)$. In-phase solutions satisfy the differential equation:

(2.4)
$$\dot{y} = g(y, \ldots, y)$$

An interesting example of a system of identical coupled oscillators is a large array of Josephson junctions that has been studied by Hadley, Beasley and Wiesenfeld [HBW1, HBW2]. The second order system of ODE for Josephson junctions is

(2.5)
$$\beta \ddot{\phi}_j + \dot{\phi}_j + \sin(\phi_j) + I_L = I_B \qquad (j = 1, \ldots N)$$

where

ϕ_j is the difference in phase of the "quasiclassical superconducting" wave functions on the two sides of the j-th junction

B is the capacitance of each junction

I_B is the bias current of the circuit

I_L is the load current.

To complete the system, assumptions must be made on how the circuit is loaded. for example, if the array is *capacitative* loaded, then

$$I_L = \sum_{j=1}^{N} \ddot{\phi}_j$$

while if the array is *resistive* loaded, then

$$I_L = \sum_{j=1}^{N} \dot{\phi}_j \ .$$

We make several observations about this system of ODE.

(a)   In-phase periodic solutions exist for a large range of the parameters β, I_B.

(b)   The Poincaré maps for the in-phase periodic solutions can loose stability by either a fixed-point or a period-doubling bifurcation - but not by a Hopf bifurcation. Both of these types of bifurcations have been found in numerical computation.

(c)   When the in-phase periodic solutions exists, it is asymptotically stable in the plane (2.3), and hence is unique.

Next we address the question of what types of solutions are expected to emanate from the bifurcations noted in (b). More detail may be found in [AGK].

Fix β, I_B at a point where an in-phase periodic solution $y(t)$ exists. Choose a Poincaré section S as follows:

(2.6) $S = L \oplus W \cong \mathbf{R}^n$

where L is the cross-section to $y(t)$ in the plane of in-phase solutions and

$$W = \{(\phi_1, \ldots, \phi_N) : \phi_1 + \cdots + \phi_N = 0\}.$$

Let $P : S \rightarrow S$ be the Poincaré map: $P(0) = 0$ since the in-phase solution is periodic.

Observe that S is S_N-invariant and that uniqueness of solutions to systems of ODE forces P to be S_N-equivariant. Indeed, we may write

(2.7) $S = L \oplus V \oplus V$

as a direct sum of S_N-irreducible subspaces, where

$$V = \{x_1 + \cdots + x_N = 0\} \cong \mathbf{R}^{N-1} \subset \mathbf{R}^N.$$

Suppose that the in-phase periodic solution is undergoing a bifurcation; that is, either of the generalized eigenspaces E_1 or E_{-1} corresponding to the eigenvalues ± 1 is nonzero. Invoking genericity, we expect the action of S_N on $E_{\pm 1}$ to be absolutely irreducible. It follows from (2.6, 2.7) that when these eigenspaces are nonzero they will generally be isomorphic to either L or V. If they happened to be isomorphic to L, then the bifurcation would produce a new in-phase periodic solution, thus contradicting (c). Hence $E_{\pm 1} \cong V$.

We consider first the possibility of a fixed-point bifurcation for the Poincaré map; that is, $E_1 \cong V$. Field and Richardson [FR] show that generically all fixed points of P have isotropy with one-dimensional fixed point subspaces. Hence the Equivariant Branching Lemma (applied to $Q(s) = P(s) - s$) implies the existence of all the expected fixed points of P. Up to permutation, the fixed points are classified as follows. Divide the oscillators into two blocks: one block having k oscillators and the other having $N - k$ oscillators. The bifurcating fixed points have the first k and the last $N - k$ coordinates equal. Their symmetry group is:

$$(2.8) \qquad\qquad \Sigma_k = S_k \times S_{N-k}.$$

Unfortunately, [IG] show that if there is a nonzero equivariant quadratic, then generally all solutions found using the Equivariant Branching Lemma are asymptotically unstable. The action of S_N on V has such a nonzero equivariant quadratic mapping.

At a period-doubling bifurcation the local bifurcation results are more interesting. Since $-I \notin S_N$ as it acts on V, we use $\widehat{S}_N$ and Theorem 2.2 to find period two solutions. There is another class of periodic solutions obtained in this way. Divide the oscillators into three blocks, the first two having k elements and the third having $N - 2k$ elements. The isotropy of such solutions is the group:

$$T_k \quad \text{generated by} \quad S_k \times S_k \times S_{N-2k} \text{ and } (x, y, z) \to -(y, x, z).$$

Theorem 2.2 implies the existence of period two points having symmetries Σ_k and symmetries T_k. The interpretation of these properties of these solutions for the Josephson function model is most interesting. As noted above, the periodic solutions with isotropy Σ_k divide the oscillators into two blocks, each block consisting of in-phase oscillators with period approximately twice that of the original in-phase solution. The periodic solutions with isotropy T_k divide the junctions into three blocks, the first two blocks consisting of in-phase oscillation but with a half period phase shift between the two blocks. The third block consists of junctions with in-phase oscillation but with a period *comparable* to the period of the in-phase periodic solution.

Certain of these solutions can be asymptotically stable (for more details, see [AGK]) and have been observed in numerical experiments on the resistive load Josephson junction model (but not with the capacitive loaded model). [AGK] also prove that there exists period two solutions to this S_N symmetric period-doubling bifurcation with submaximal isotropy. For these solutions the oscillators divide into three blocks of unequal size.

Lecture 3: *Genericity and Boundary Conditions*

The Faraday experiment provides an example where a general qualitative analysis based on period-doubling bifurcations may connect theory with experiment. This connection highlights the effects that boundary conditions may have on genericity (see [CGGKS] where the issues raised here are discussed more fully).

In the Faraday experiment a fluid layer is subjected to a vertical oscillation at frequency ω and forcing amplitude A. When A is small the fluid remains essentially flat and when A is increased the flat surface bifurcates to a standing wave at frequency $\omega/2$. What is measured in the experiments of Gollub and coworkers [CG, GS] is a *stroboscopic map* S which pictures the surface of the fluid at each period of the forcing. Since, after the bifurcation, the fluid surface returns to its original form each second iterate of S, we have a period-doubling bifurcation.

The experiments of Ciliberto and Gollub [CG] focus on fluid layers with circular cross-section, while the experiments of Gollub and Simonelli [GS] focus on the square cross-section case. A qualitative analysis of the circular cross-section experiment, along the lines that we describe here for the square cross-section, is given by Crawford, Knobloch and Riecke [CKR].

The following points are observed in the experiments [GS].

(a)   For most values of the forcing frequency ω the initial bifurcation from stability of the flat surface as the amplitude A is increased is by a period-doubling bifurcation. Spatial modes are detected and described by their wave numbers in both horizontal directions, such as (3,1), (3,2), (4,0).

(b)   For isolated values of ω the flat surface loses stability to two modes simultaneously.

Given this information we may make several reasonable assumptions concerning the mathematical analysis of any model purporting to describe the Faraday experiment.

(a)   Since The experiment is square symmetric, the loss of stability of the flat surface to say a (3,1) mode would imply loss of stability to the (1,3) mode as well; that is, the eigenspace E_{-1} corresponding to the period-doubling -1 eigenvalue is at least double. Generically, it is precisely double, and hence $E_{-1} \cong \mathbb{C}$.

(b)   At the mode interaction point E_{-1} is isomorphic to $\mathbb{C}^2$.

(c)   Assuming that a center manifold reduction is possible, the dynamics of the (stroboscopic map of the) Faraday experiment near the mode interaction point is controlled by the dynamics of a D_4-equivariant mapping $f : \mathbb{C}^2 \to \mathbb{C}^2$.

These assumptions, however, lead to a difficulty. The representation of the symmetry group of the square, D_4, on the eigenspace E_{-1} at a generic (non-mode-interaction point) is either an irreducible two-dimensional representation of D_4 or the sum of two one-dimensional irreducibles.

In the latter case we must then question why a nongeneric situation occurs in this experiment (since generically eigenspaces are irreducible). In the former case the representation is irreducible, but a different problem occurs at a point of mode interaction. Up to isomorphism the two-dimensional irreducible representation of D_4 is unique. Hence, at a codimension two point of mode interaction of two two-dimensional, isomorphic, irreducible representations, generically we expect the linearization (of f) to be nilpotent. This nilpotency would imply that there is only one independent set of eigenfunctions (not two), and hence that the two distinct modes would, in fact, have to merge together at the codimension two point (and be physically indistinguishable).

We are faced with a dilemma: either something nongeneric (the reducibility of the eigenspace) occurs in models of the Faraday experiment, or something is wrong with the experimental observation of distinct modes in the square cross-section case.

We present here an alternative explanation based on some subtleties of genericity and boundary conditions. See [CGGKS].

Fujii, Mimura and Nishiura [FMN] and Armbruster and Dangelmayr [AD] observed that the bifurcation of steady solutions in reaction-diffusion equations on the line changed from what might have been expected when Neumann boundary conditions (NBC) were assumed. We abstract part of their reasoning here.

Any solution u to a reaction-diffusion equation on $[0, \pi]$ with NBC can be extended in a solution v to that same equation with periodic boundary conditions (PBC) on $[-\pi, \pi]$ by extending the solution to be even across zero. More precisely, define:

$$(3.1) \qquad v(x,t) = u(-x,t) \quad \text{for all} \quad x < 0.$$

Conversely solutions v to the PBC problem that are also even (which, using (3.1) is a fixed point subspace condition for the symmetry $x \to -x$) is a solution to the NBC model.

What is gained by the extension to PBC is the introduction of $O(2)$ symmetry into the problem (translational symmetry of the reaction-diffusion equation modulo the 2π periodicity of the boundary conditions). The idea for determining genericity is to look at the generic PBC case (that is, $O(2)$ symmetric bifurcation) and then restrict (by fixed-point subspace arguments) to the NBC case.

Similar statements about genericity are valid for Dirichlet boundary conditions (DBC), although DBC does require an extra reflectional symmetry on the differential operator to be valid in order to make the extension to PBC. This symmetry is valid, for example, in the Navier–Stokes equations. Indeed, similar statements hold for systems and for higher dimensional domains with various mixtures of boundary conditions.

We now return to the Faraday experiment. In any analytic model of the experiment one must solve for both the *surface deformation* $\zeta(x, y)$ and the *fluid velocity field* $u(x, y, z)$. Typically, in models, no-slip or Dirichlet boundary conditions are valid for u along the lateral boundaries and Neumann boundary conditions are assumed on ζ (that is, the fluid surface is assumed to be perpendicular to the side walls).

As our discussion above indicates these boundary conditions have the effect of introducing T^2 symmetry into the bifurcation problem. The two-torus T^2 is obtained by planar translations modulo the double periodicity of the square. Thus, the full symmetry group Γ of the Faraday experiment with square geometry is generated by D_4 and T^2.

The consequence of having this enlarged symmetry group is that all of the two-dimensional eigenspaces noted above are irreducible representations of Γ, and that, at mode interaction points, distinct modes have distinct irreducible representations. Hence, the linearization is diagonal rather than nilpotent and distinct modes need not merge (thus agreeing with experimental observation).

Period-doubling bifurcations at points of mode interaction in the Faraday experiment is being studied in [CGK] using the group $\widehat{\Gamma}$.

Acknowledgement

The research described in these lectures was supported in part by the Institute for Mathematics and its Applications, University of Minnesota, and by the following research grants: NSF/Darpa (DMS-8700897), Texas Advanced Research Program (ARP-1100) and NASA-Ames (NAG2-432).

Many of the results discussed in these lectures were the product of unpublished collaborative research. I wish to thank my collaborators on these projects for permitting me free use of this material: Don Aronson, John David Crawford, Gabriella Gomes, Edgar Knobloch, Maciej Krupa, and Ian Stewart.

REFERENCES

[AD] D. ARMBRUSTER & G. DANGELMAYR, *Coupled stationary bifurcations in non-flux boundary value problems*, Math. Proc. Comb. Phil. Soc. 101 (1987), 167–192.

[AGH] D. ARMBRUSTER, J. GUCKENHEIMER & P. HOLMES, *Heteroclinic cycles and modulated travelling waves in systems with 0(2) symmetry*, Physica 29 D (1988) 257–282.

[AGK] D.G. ARONSON, M. GOLUBITSKY & M. KRUPA, *Coupled arrays of Josephson junctions and bifurcation of maps with S_N symmetry*, Nonlinearity. Submitted.

[BH] F.H. BUSSE & K.E. HEIKES, *Convection in a rotating layer: a simple case of turbulence*, Science 208 (1980) 173–175.

[ChG] P. CHOSSAT & M. GOLUBITSKY, *Symmetry-increasing bifurcation of chaotic attractors*, Physica 32 D (1988) 423–436.

[C] G. CICOGNA, *Symmetry breakdown from bifurcations*, Lettere al Nuovo Cimento 31 (1981) 600–602.

[CG] S. CILIBERTO & J. GOLLUB, *Chaotic mode competition in parametrically forced surface waves*, J. Fluid Mech. 158 (1985), 381–398.

[CGGKS] J.D. CRAWFORD, M. GOLUBITSKY, M.G.M. GOMES, E. KNOBLOCH & I.N. STEWART, *Boundary conditions as symmetry constraints*, Preprint, University of Warwick (1989).

[CKR] J.D. CRAWFORD, E. KNOBLOCH & H. RIECKE, *Competing parametric instabilities with circular symmetry*, Phys. Lett. A 135 (1989) 20–24.

[F1] M. FIELD, *Equivariant dynamical systems*, Trans. A.M.S. 259, No. 1 (1980) 185–205.

[F2] M. FIELD, *Equivariant bifurcation theory and symmetry breaking*, Dyn. Diff. Eqn. 1, No. 4 (1989), 369–421.

[FR1] M. FIELD & R.W. RICHARDSON, *Symmetry breaking and the maximal isotropy subgroup conjecture for reflection groups*, Arch. Rational Mech. Anal. 105 (1989) 61–94.

[FR2] M. FIELD & R.W. RICHARDSON, *New examples of symmetry breaking and the distribution of symmetry breaking isotropy types*, In preparation.

[FMN] H. FUJII, M. MIMURA & Y. NISHIURA, *A picture of the global bifurcation diagram in ecologically interacting and diffusing systems*, Physica 5D (1982) 1–42.

[GH1] J. GUCKENHEIMER & P. HOLMES, *Nonlinear Oscillations, Dynamical Systems, and bifurcation of Vector Fields,*, Appl. Math. Sci. 42, Springer-Verlag, New York, 1983.

[GH2] J. GUCKENHEIMER & P. HOLMES, *Structurally stable heteroclinic cycles*, Math. Proc. Comb. Phil. Soc. 103, part 1 (1988), 189–192.

[GSS] M. GOLUBITSKY, I.N. STEWART & D.G. SCHAEFFER, *Singularities and Groups in Bifurcation Theory: Vol. II*, Applied Math. Sci. 69, Springer-Verlag, New York, 1988.

[HBW1] P. HADLEY, M.R. BEASLEY & K. WIESENFELD, *Phase locking of Josephson-junction series arrays*, Phys. Rev. B 38, No. 13 (1988) 8712–8719.

[HBW2] P. HADLEY, M.R. BEASLEY & K. WIESENFELD, *Phase locking of Josephson-junction arrays*, Applied Phys. Lett. 52, No. 19 (1988), 1619–1621.

[IG] E. IHRIG & M. GOLUBITSKY, *Pattern section with 0(3) symmetry*, Physica 12D (1984), 1–33.

[ML] R.M. MAY & W.J. LEONARD, *Nonlinear aspects of competition between three species*, SIAM J. Appl. Math. 29 (1975), 243–253.

[M] I. MELBOURNE, *Intermittancy as a codimension three phenomenon*, Dyn. Diff. Eqn. 1, No. 4 (1989), 347–368.

[MCG] I. MELBOURNE, P. CHOSSAT & M. GOLUBITSKY, *Heteroclinic cycles involving periodic solutions in mode interactions with O(2) symmetry*, Proc. R. Soc. Edinburgh 113A (1989), 315–345.

[R] D. RUELLE, *Bifurcations in the presence of a symmetry group*, Arch. Rational Mech. Anal. 51 (1973), 136–152.

[SG] F. SIMONELLI & J. GOLLUB, *Surface wave mode interactions: effects of symmetry and degeneracy*, J. Fluid Mech. 199 (1989), 471–494.

[V] A. VANDERBAUWHEDE, *Local Bifurcation and Symmetry*, Habilitation Thesis, Rijksuniversiteit Ghent, 1980; Res. Notes in Math. 75, Pitman, Boston, 1982.

DYNAMICS OF SOME ELECTROCHEMICAL REACTIONS

J. L. HUDSON*

Abstract. Experiments on a few electrochemical reactions are discussed. Time series of either current or voltage, obtained under potentiostatic or galvanostatic conditions respectively, are presented and characterized. We first some some examples of dynamic behavior such as chaos, quasiperiodicity, and period doubling of tori obtained during the electrodissolution of copper. Some apparent higher order chaos during electrodissolution of iron is then discussed. Finally, we treat briefly coupled electrochemical oscillators.

Key words. dynamics, chaos, electrochemical reactions

1. Introduction. Chemically reacting systems are often highly nonlinear and thus have furnished many examples of interesting dynamic behavior. For example, the exothermic, single irreversible reaction (a two variable system) has been analyzed extensively by chemical engineers; the existence of multiple states and oscillatory behavior is now well understood [2,3,4,22,40,41]. Two independent reactions, such as consecutive reactions, with heat effects are governed by three ODE's and can, in addition, produce chaos [16,17,19,20].

There has been a considerable amount of experimental work done on the dynamics of isothermal reactions carried out in continuous stirred reactors. In these cases the nonlinearities are associated with the chemical kinetics themselves rather than with the Arrhenius temperature dependence of the reaction rate constant as in the nonisothermal reactions discussed above [11,15]. The best known isothermal oscillating reaction system is the Belousov-Zhabotinski reaction. Much of the early experimental work on chemical chaos was done with this reaction [18,31,34].

In this paper we discuss the dynamics, with emphasis on chaotic behavior, of some electrochemical reactions. Experiments with electrochemical systems have produced examples of several types of periodic, quasiperiodic, and low order chaotic behavior. [1,6–10, 25,29 37,38]. In most cases the dynamics are fast (with frequencies much higher than those obtained in stirred reactors) and reasonably stationary so that analyses of the time series can be carried out by standard methods such as attractor reconstruction, dimension calculations, construction of return maps and Poincaré sections, etc.

We concentrate on the dynamics of the reactions (rather than on the physiochemical behavior of the electrode surface or the possible causes of instabilities). Two reactions are discussed: The electrodissolution of copper and that of iron.

*Department of Chemical Engineering, University of Virginia, Charlottesville, VA 22903-2442.
This work was supported in part by the National Science Foundation and by the Center for Innovative Technology, Commonwealth of Virginia.

2. Experiments. All of the experiments were carried out with a standard electrochemical apparatus, viz., a rotating disk electrode. This is a rod of the appropriate metal, exposed at one end, which is rotated in an electrolyte. The overall rate of reaction is influenced by the kinetics on the electrode surface, by transport through any film which may exist on the electrode surface, and by mass transfer in the solution; the latter can be controlled by the rotation rate of the disk.

Experiments are normally carried out either potentiostatically or galvanostatically. In the former case current is measured as a function of time; in the latter case the potential relative to some reference is measured. Data are taken digitally at rates fast enough to analyze the time signal. In the experiments discussed here this is in the range of 60 to 5000 Hz.

3. Copper electrodissolution. Potentiostatic oscillations have been observed using a copper electrode in solutions of sulfuric acid and sodium chloride [6–9]. Under some conditions the current undergoes temporal oscillations. The exact mechanism of the reaction and the cause of the oscillations have not yet been determined. Some of the reactions which occur under the conditions of the experiment, however, are the following: The copper is oxidized via the anodic reaction [23,24].

$$Cu + Cl^- \rightarrow CuCl + e^-.$$

The CuCl forms a film on the surface of the copper electrode and this film serves as a resistance to the transport of Cl^- ions. The film dissolves via the reactions

$$CuCl + (x-1)\ Cl^- \rightarrow CuCl_x^{1-x}\ (aq)$$

and [23,24]

$$CuCl \rightarrow Cu^{+2} + Cl^- + e^-.$$

The rate of formation of the film is somewhat greater than the rate of dissolution, and there is thus a slow net growth in the thickness of the CuCl film, and therefore also in the resistance to transport of Cl^- to the metal surface. This resistance appears to be an important parameter for the system. In some of the results of this section transitions among states are shown; these transitions are caused by changes in a slowly varying parameter, viz., the properties and thickness of the layer on the disk surface.

Evidence for several types of low-dimensional chaotic behavior has been obtained with this reaction system. We show a few examples. In Figure 1 a time series exhibiting chaotic features is shown [9]. The attractor (Figure 1b), Poincaré section (Figure 1c), and return map (Figure 1d) indicate that the time series (Figure 1a) is simple chaos. A sequence of Poincaré sections is shown in Figure 1e to explain the flow of trajectories around the attractor. The Poincaré sections were made with planes at constant $I(t - 2\tau) = 17.5\ mA$ and $I(t - 4\tau) = 17.5\ mA$ and were constructed by using all the intersections of the attractor of Figure 1b with

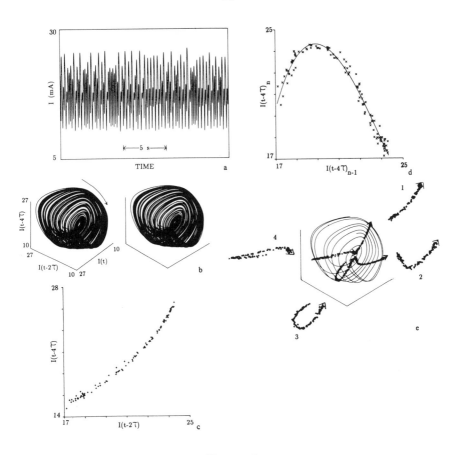

FIGURE 1.

chaos: (a) time series. (b) Attractor (spline fit). (c) Poincaré section taken at $I(t) = 17.5\ mA$; increasing $I(t - 4\tau)$. (d) Return map constructed from the Poincaré section taken at $It - 2\tau) = 17.5\ mA$ with increasing $I(t - 4\tau)$; using variable $I(t - 4\tau)$. (e) Series of Poincaré sections showing the flow of trajectories around the attractor. Sections 1 and 3 were taken at $I(t - 2\tau) = 17.5\ mA$ and sections 2 and 4 were taken at $I(t - 4\tau) = 17.5\ mA$ [9].

the above-mentioned planes. In Figure 1e only a few orbits around the attractor are shown so that the location and orientation of the section can be seen. The arrow on each section is used to mark the points that started on the edge of the attractor in section 1. This sequence shows the stretching, folding and mixing, and contraction processes common to these types of simple chaotic attractors. In going from section

1 to section 2, one observes the stretching of the trajectories. In going from section 2 to 3, the trajectories continue to stretch but the folding process is also observed. In addition, the points that were on the outside of the attractor are now on the inside. In going from section 3 to 4, one sees the contraction of the fold onto itself. This process continues as trajectories proceed around the attractor to give back the result seen in section 1.

From the shapes of the attractor, Poincaré sections, and return map, the chaos shown in Figure 1 appears to be a simple type that may be embeddable in three-dimensional space and that has a single positive Lyapunov exponent. For further verification the correlation dimension [14] and the largest Lyapunov exponent [35] were calculated. The correlation dimension was found to be 2.2. An approximate value of 0.6 for the Lyapunov exponent was calculated from the return map.

The chaotic behavior shown in Figure 1 was preceded by a period doubling of periodic orbits and chaos on a two-band attractor.

Another type of behavior, apparently chaos on a broken toroidal structure, is shown in Figure 2. The attractor is shown in Figure 2b, and a Poincaré section, a closed curve indicating a toroidal substructure, is shown in Figure 2c. A series of Poincaré sections (Figure 2d) shows the stretching and folding process that indicates the chaotic nature of flow. The planes used to make the sections are defined by the equation $I(t)+I(t-2\tau)-40 = \alpha(I(t-4\tau)-20)$. Changing the value α rotates the plane about a line that goes through the center of the attractor and is defined by the intersection of the planes $I(t)+I(t-2\tau) = 40$ and $I(t-4\tau) = 20$. The first cut shows a closed cross section. Two areas of the section are marked (A and B) to point out the folding process. The folding become noticeable in the second section as A and B are coming together. The folding process is almost complete in the third section as A and B have almost totally come together. By the fourth section the folds have become so close together that the separation can no longer be seen. The fifth section is $180°$ from section one; in the remaining $180°$ the shapes of the sections undergo no additional significant changes. The behavior shown in Figure 2 was followed by quasiperiodic behavior which is not shown.

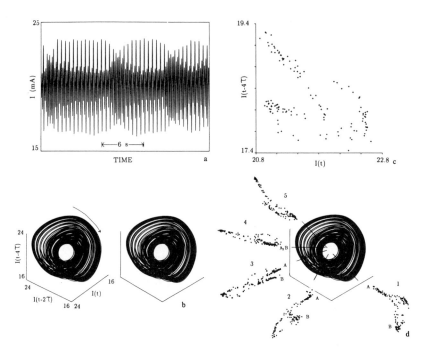

FIGURE 2.

Chaos on a broken torodial structure: (a) Time series. (b) Attractor (spline fit). (c) Poincaré section taken at $I(t - 2\tau) = 20 \ mA$; increasing $I(t - 2\tau)$. (d) Series of Poincaré sections showing the folding process. All planes used to make the sections are defined by the general equation $I(t)+I(t - 2\tau) - 40 = \alpha(I(t - 4\tau) - 20)$. For section 1 and 5, $\alpha = 1$; for section 2, $\alpha = -1$; for section 3, $\alpha = -5$; and for section 4, $\alpha = 7$. These planes are marked by lines on the attractor. Attractor is the same as that shown in Figure 2b [9].

We do show a different example of quasiperiodic behavior which was obtained during a sequence of period halving of tori [8]. The observed sequence was chaos on a broken torus, chaos on a two-band broken torus, double torus, torus, limit cycle, and steady state. Each of these transitions occurred smoothly. For example, the cross section of the torus decreased slowly until the limit cycle resulted; the amplitude of the limit cycle decreased resulting on a steady state. The flow on a torus is shown on Figure 3. The time series (Figure 3a) has quasiperiodic features; there are oscillations of period 0.76 s, the maxima of which are also periodic, with a period of 3.57 s. The power spectrum (figure 3b) has two incommensurate frequencies $f_1 = 1.32$ Hz and $f_2 = 0.28$ Hz both of which have peaks well above the underlying

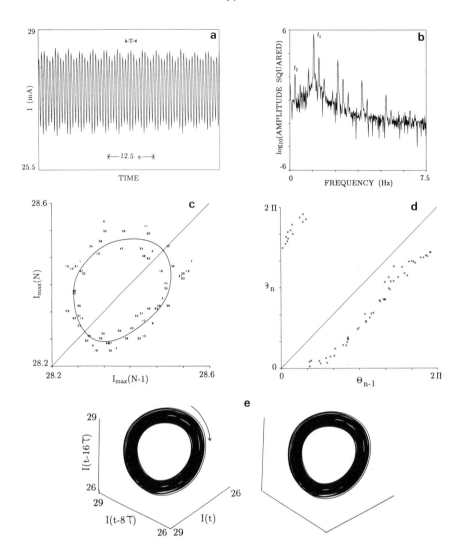

FIGURE 3.

Torus. (a) Time series. (b) Power spectrum; $f_1 = 1.32$ Hz; $f_2 = 0.28$ Hz. (c) Next maximum map; (d) Next angle return map of (c); center point used for calculation was (28.38, 28.38). (e) Attractor (spline fit); the two projections are made at angles differing by 6°. The right picture is for the left eye and the left for the right. ($\tau = 1/60$s) [8].

noise. The frequency f_1 is associated with the flow around the main loops of the torus. The frequency $f_2(0.28$ Hz$)$ is associated with progression in the plane of the Poincaré section or the next maximum map (Figure 3c). This progression is in a clockwise direction in the next maximum map. The time associated with this progression, $T = 1/f_2 = 3.57$ s, is also indicated on the time series. Both the Poincaré section (not shown) and the next maximum map (Figure 3c) should be closed curves for quasiperiodic behavior. The next maximum map is shown here rather than a Poincaré section since, as is usually the case with experimental data, it has less scatter.

A next angle return map was made from Figure 3c and is shown in Figure 3d. This map was made by first translating the coordinate system of Figure 3c to (28.38, 28.38) which lies in the hole of the map. The coordinate system was then rotated so that the first point of the next maximum map was at 0°. The angle of the $n+1$st point with respect to this new coordinate system was plotted against the angle of the nth point for all the points in the map. Figure 3d appears to be approximately one-dimensional and invertible indicating quasiperiodic behavior. A reconstruction of the attractor is presented in Figure 3e and has the appearance of a torus.

The quasiperiodic behavior of Figure 3 was preceded by the quasiperiodic behavior of figure 4. As can be seen from the time series in Figure 4a, the maxima of the oscillations now form an oscillatory pattern whose period is doubled when compared to that of Figure 3a. The power spectrum (Figure 4b) shows two incommensurate, fundamental frequencies at $f_1 = 1.32$ Hz and $f_2/2 = 0.14$ Hz. The frequency f_1 corresponds to the main flow around the center hole of the torus. The frequency $f_2/2$ represents progression in the plane of a Poincaré section (not shown) or equivalently in the next maximum map shown in Figure 4c. The frequency doubling which has occurred in going from Figure 3 to Figure 4 is associated with the frequency of the progression in the plane of the Poincaré section or in the next maximum map.

The frequency $f_2/2$ can be seen in the power spectrum, Figure 4b. Although the frequency $f_2/2$ is not overly strong, it does not appear to be part of the underlying noise. This argument is further supported by the following two observations. First, the frequency $f_2/2$ is followed in the power spectrum by the integer multiples $2(f_2/2)$, $3(f_2/2)$, and $4(f_2/2)$. Secondly, we can estimate the frequency $f_2/2$ directly from the time series and compare the result favorably to $f_2/2 = 0.14$ Hz. The time corresponding to frequency $f_2/2 = 0.14$ s, is indicated by the time segment $T = T_L + T_s$ on fig. 4a. The time T_L represents the transversal of the larger loop of Figure 4c and T_s represents the transversal of the smaller. Thus the wave forms associated with the maxima of the oscillations as seen in Figure 4a contain a larger maxima segment for T_L and a smaller maxima segment for T_s, and the total is commensurate with the frequency $f_2/2$.

The next angle return map is shown in Figure 4d. The coordinates now go up to 4π since there is a double loop in Figure 4c.. Figure 4d was constructed in the same manner as the next angle return map in Figure 3d; a point was chosen inside the small loop of Figure 4c and the angles calculated from it. A reconstruction of the attractor is shown in Figure 4e.

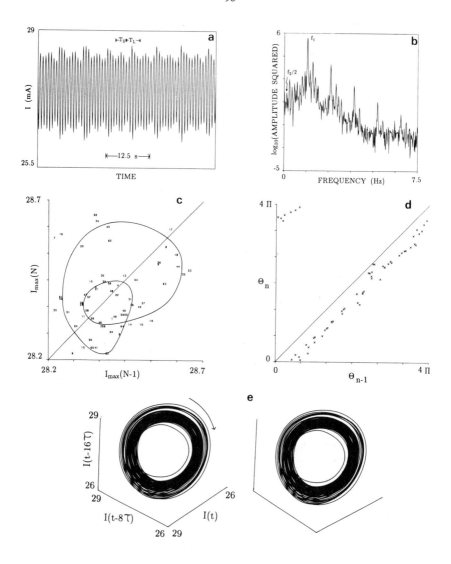

FIGURE 4.

Double torus. (a) Time series. (b) Power spectrum; $f_1 = 1.32$ Hz; $f_2/2 = 0.14$ Hz. (c) Next maximum map. (d) Next angle return map of (c); center point used for the calculation was (28.38, 28.38). (e) Attractor (spline fit) [8].

The double torus of Figure 4 is preceded by what appears to be chaos on a broken double toroidal structure and that itself preceded by chaos on a broken torus.

4. Iron electrodissolution. Oscillations in current also occur during the potentiostatic electrodissolution of iron in acidic solutions [10,13, 32]. Those oscillations which occur on the mass transfer limited plateau can be faster and more complicated than those seen above in the previous section.

An example of a time series for the iron/sulfuric acid system is shown in Figure 5. Note that only one second of data is shown. These data were taken at 5000 Hz.

The detailed nature of the time series depends, of course, on parameters such as acid concentration, disk rotation rate, applied potential, and the size of the disk. We consider here only the latter.

The complexity of the time series increases with increasing electrode size. For example, increasing the disk diameter in small increments from 2.0 mm to 6.25 mm yields a series of behaviors from a limit cycle to low order chaos (correlation dimension slightly over 2.0) to a higher order chaos (correlation dimension well above 3.0). Figure 5 shows the time series for the largest electrode.

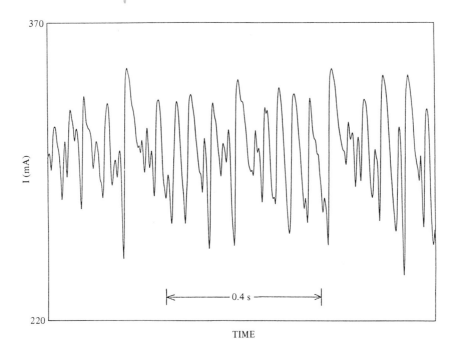

FIGURE 5.

Time series from iron/sulfuric acid system

We do not know why the complexity increases with an increase in the area of the electrode. One obvious possibility is the increase in the number of reacting sites on the surface.

5. Coupled electrodes. The final comment of the last section leads perhaps naturally to the subject of coupled electrochemical reactors. The subject of coupled chemical oscillators has received considerable, numerical and analytical attention [5,21,26,27,30,39]. However, there are few experimental studies [36]. It is difficult to couple chemical reactors without changing the nature of the uncoupled oscillators in the process.

Electrodes may be suitable for studies of coupled chemical oscillators. We have carried out some preliminary experiments with two electrodes. One was a rotating disk electrode similar to those used in the experiments described in the preceding sections. The second was a similar, but non-rotating, electrode placed opposite the rotating electrode on the same axis. Two potentiostats and measurement systems are used so that all the coupling occurs through the electrolyte. The experiments are run under conditions such that each electrode would oscillate if the other were not active. In our preliminary experiments only frequency locking has been observed. We hope to exploit this system (or most likely other geometries) to investigate other phenomena of coupled chemical oscillators.

6. Final Comments. We have shown some dynamic behavior obtained with the studies on the electrodissolution of metals. Other types of electrochemical reactions have received less attention from a dynamic standpoint; however, some work has been done recently on electrodeposition and electrocatalytic reactions [12,33].

We note that several types of periodic and low dimensional chaotic behavior have been observed. Although higher order chaos can occur, as discussed briefly on Section 4, it is interesting that such a degree of order can prevail in such complicated systems.

REFERENCES

[1] F. N. ALBAHADILY AND M. SCHELL, *An experimental Investigation of Periodic and Chaotic Electrochemical Oscillations in the Anodic Dissolution of Copper in Phosphoric Acid*, J. Chem. Phys. 88 (1988), pp. 4312–4319..

[2] R. ARIS, *The Mathematical Background of Chemical Reactor Analysis. II. The Stirred Tank*, in Reacting Flows, G.S.S. Luddford (ed.), Lectures in Applied Math, 24 (1986), pp. 75–108.

[3] R. ARIS, AND N. R. AMUNDSON, *An Analysis of Chemical Reactor Stability and Control-I*, Chem. Eng. Sci., 7 (1958), pp. 121–131.

[4] V. BALAKOTAIAH AND D. LUSS, *Analysis of Multiplicity Patterns of a CSTR*, Chem. Eng. Commun., 13 (1981), pp. 111–132.

[5] K. BAR-ELI, *On the Coupling of Chemical Oscillators*, J. Phys. Chem., 88 (1984), pp. 3616–3622.

[6] M. R. BASSETT AND J. L. HUDSON, *Experimental Evidence for Period Doubling of Tori During an Electrochemical Reaction*, Physica D, 35 (1989), pp. 289–298.

[7] M. R. BASSETT AND J. L. HUDSON, *The Dynamics of the Electrodissolution of Copper*, Chem. Eng. Comm., 60 (1987), pp. 145–159.

[8] M. R. BASSETT AND J. L. HUDSON, *Shil'nikov Chaos During Copper Electrodissolution*, Journal of Physical Chemistry, 92 (1988), pp. 6963–6966.

[9]   M. R. BASSETT AND J. L. HUDSON, *Quasiperiodicity and Chaos During An Electrochemical Reaction*, Journal of Physical Chemistry, 93 (1989), pp. 2731–2737.

[10] C. B. DIEM AND J. L. HUDSON, *Chaos During the Electrodissolution of Iron*, A.I.Ch.E. J., 33 (1987), pp. 218–224.

[11] M. FEINBERG, *Reaction Network Structure, Multiple Steady States, and Sustained Composition Oscillations: A Review of Some Results*, in Modelling of Chemical Reaction Systems,, K. H. Ebert, et al. eds., Springer, Berlin (1981).

[12] N. FETNER AND J. L. HUDSON, *Oscillations during the electrocatalytic reduction of hydrogen peroxide on a platinum electrode*, preprint, (1990).

[13] U. F. FRANCK AND R. FITZHUGH, *Periodische Elektrodenprozesse und ihre Beschreibung durch ein mathematisches Modell*, Z. Electrochem, 65 (1961), pp. 156–168.

[14] P. GRASSBERGER AND I. PROCACCIA, *Characterization of Strange Attractors*, Phys. Rev. Lett., 50 (1983), pp. 346–349.

[15] P. GRAY AND S. K. SCOTT, *Autocatalytic Reactions in the Isothermal CSTR: Oscillations and Instabilities in the system $A + 2B \rightarrow 3B$, $B \rightarrow C$*, Chem. Eng. Sci., 39 (1984), pp. 1087–1097.

[16] J. L. HUDSON AND I. G. KEVREKIDIS, *Dynamic Behavior or Nonisothermal Continuous Stirred Tank Reactors* in Handbook of Heat and Mass Transfer, N. P. Cheremisinoff, ed., Gulf, Houston, 3 (1989), pp..

[17] J. L. HUDSON AND O. E. ROSSLER, *Dynamics of Nonlinear Systems*, V. Hlavacek, ed., Gordon and Breach Science: New York (1986), Chapter 6.

[18] J. L. HUDSON, M. HART, AND D. MARINKO, *An experimental study of multiple peak periodic and nonperiodic oscillations in the Belousov-Zhabotinskii reaction*, J. Chem. Phys. 71 (1979), pp. 1601–1606.

[19] D.V. JORGENSEN AND R. ARIS, *On the Dynamics of a Stirred Reactor with Consecutive Reactions*, Chem. Eng. Sci., 38 (1983), pp. 45–53.

[20] C. KAHLERT, O. E. ROSSLER, AND A. VARMA, *Chaos in a Continuous Stirred Reactor with Two Consecutive First Order Reactions, one Exo–, One Endothermic*, in Modeling Chemical Reaction Systems, K. Ebert and W. Jaeger, ed., Springer, Heidelberg (1981).

[21] C. R. KENNEDY AND R. ARIS, *Bifurcations of a Model Diffusion–Reaction System*, in New Approaches to Nonlinear Problems in Dynamics, P. J. Holmes, ed., SIAM (1980).

[22] V. K. KWONG AND T. T. TSOTSIS, *Fine Structure of the CSTR Parameter Space*, AIChE J., 29 (1983), pp. 343–347.

[23] H. P. LEE AND K. NOBE, *Kinetics and Mechanisms of Cu Electrodissolution in Chloride Media*, J. Electrochem, Soc., 133 (1986), pp. 2035–2043.

[24] H. P. LEE, K. NOBE, AND A. PEARLSTEIN, *Film Formation and Current Oscillations in the Electrodissolution of Cu in Acidic Chloride Media*, I. Experimental Studies, J. Electrochem. Soc., 132 (1985), pp. 1031–1037.

[25] O. LEV, A. WOLFFBERG, M. SHEINTUCH, AND L. M. PISMEN, *Bifurcations to Periodic and Chaotic Motions in Anodic Nickel Dissolution*, Chem. Eng. Sci., 43 (1988), pp. 1339–1353.

[26] J. C. MANKIN AND J.L. HUDSON, *The dynamics of Coupled Nonisothermal Continuous Stirred Tank Reactors*, Chem. Eng. Sci., 41 (1986), pp. 2651–2661.

[27] H. G. OTHMER, D. G. ARONSON, AND E. J. DOEDEL, RESONANCE AND BISTABILITY IN COUPLED OSCILLATORS, Phys. Letters, 113A (1986), pp. 349–353.

[28] N. H. PACKARD, J. P. CRUTCHFIELD, J. D. FARMER, AND R. S. SHAW, *Geometry from a Time Series*, Phys. Rev. Lett., 45 (1980), p. 712.

[29] A. J. PEARLSTEIN, H. P. LEE, AND K. NOBE, *Film Formation and Current Oscillations in the Electrodissolution of Copper in Acidic Chloride Media*, II. Mathematical Model, J. Electrochem. Soc., 132 (1985), pp. 2159–2165.

[30] V. RAVI KUMAR, B. JAYARAMAN, B. KULKARNI, AND L. DORAISWAMY, *Dynamic Behavior of Coupled CSTRs Operating Under Different Conditions*, Chem. Engng. Sci., 38 (1983), pp. 673–686.

[31] J. C. ROUX, J. S. TURNER, W. D. MCCORMICK, AND H. L. SWINNEY, *Experimental observations of complex dynamics in a chemical reactor*, in Nonlinear Problems Present and Future, A. R. Bishopp, ed. North Holland (1981).

[32] P. RUSSELL AND J. NEWMAN, *Current Oscillations Observed within the Limiting Current Plateau for Iron in Sulfuric Acid*, J. Electrochem. Soc., 133 (1986), pp. 2093–2097.

[33] M. SCHELL, F. N. ALBAHADILY, J. SAFIR, AND Y. YU, *Characterization of oscillatory states in the electrochemical oxidation of formaldehyde and formate/formic acid*, J. Phys. Chem., 83 (1989), pp. 4806–4810.

[34] R. A. SCHMITZ, R. R. GRAZIANI, AND J. L. HUDSON, *Experimental evidence of chaotic states in the Belousov-Zhabotinskii reaction*, J. Chem. Phys. 67 (1977), pp. 3040–3044.

[35] H. G. SCHUSTER, *Deterministic Chaos*, Physik-Verlag: Weinheim (1984).

[36] I. STUCHL AND M. MAREK, *Dissipative Structures in Coupled Cells: Experiments*, J. Chem. Phys. 77 (1982), pp. 2956–2963.

[37] J. B. TALBOT AND R. A. ORIANI, *Application of Linear Stability Analysis to Passivation Models*, J. Electrochem. Soc., 132 (1985), pp. 1031–1037.

[38] L. T. TSITSOPOULOS, T. T. TSOTSIS, AND I. A. WEBSTER, *An Ellipsometric Investigation of Reaction Rate Oscillations During the Electrochemical Anodization of Cu in H_3PO_4 Solutions*, Surface Science, 191 (1987), pp. 225–238.

[39] T. T. TSOTSIS, *Spatially Patterned States in Systems of Interacting Lumped Reactors*, Chem. Eng. Sci., 38 (1983), pp. 701–717.

[40] A. UPPAL, W. H. RAY, AND A. POORE, *On the Dynamic Behavior of Continuous Stirred Tank Reactors*, Chem. Eng. Sci., 29 (1974), pp. 967–985.

[41] A. VARMA AND R. ARIS, *Stirred Pots and Empty Tubes*, in Chemical Reactor Theory: A Review, L. Lapidus and N. R. Amundson (ed.), Prentice-Hall, (1977), 79–155..

CONSTRUCTION OF THE FITZHUGH-NAGUMO PULSE USING DIFFERENTIAL FORMS

C. JONES*, N. KOPELL** AND R. LANGER†

1. Introduction. Systems of singularly perturbed ordinary differential equations can often be solved approximately by singular solutions. These singular solutions are pieced together from solutions to simpler sets of equations obtained as limits from the original equations. There is a large body of literature on the question of when the existence of a singular solution implies the existence of an actual solution to the original equations. Techniques that have been used include fixed point arguments (Conley [1], Carpenter [2], Hastings [3] and Gardner and Smoller [4]), implicit function theorem and related functional-analytic techniques (Fife [5], Fujii et al. [6], Hale and Sakamoto [7,8]) differential inequalities (see for instance Chang and Howes [9]) and nonstandard analysis (Diener and Reeb [10]).

Some of these techniques apply to nerve impulse equations, including the Hodgkin-Huxley equations and a widely used simplified version known as the FitzHugh-Nagumo equations. These equations are

$$\begin{aligned}\frac{\partial u}{\partial t} &= \frac{\partial^2 u}{\partial x^2} + f(u) - w \\ \frac{\partial w}{\partial t} &= \epsilon(u - \gamma w)\end{aligned}$$

(1.1)

where ϵ is a small parameter and $f(u) = u(1 - u)(u - a)$, with a any constant satisfying $0 < a < 1/2$. The solutions of interest are *travelling pulses*. These are solutions depending only on $\zeta = x + \theta t$, for some wave speed θ and which, as $\zeta \to \infty$ or $\zeta \to -\infty$, approach the unique rest state of (1.1) (γ is chosen small enough so that $u = \gamma w$ and $w = f(u)$ have only (0,0) as a simultaneous solution). These travelling waves then satisfy

(1.2)
$$\begin{aligned}u' &= v \\ v' &= \theta v - f(u) + w \qquad\qquad ' = d/d\zeta \\ w' &= \frac{\epsilon}{\theta}(u - \gamma w)\end{aligned}$$

together with the conditions $(u, v, w) \to (0, 0, 0)$ as $\zeta \to \pm\infty$. It was proved independently by Carpenter [2] and Hastings [3] that there is a value of θ at which a travelling pulse exists, if ϵ is sufficiently small. In fact there are two values of θ but we are interested in the faster pulse.

---

*Department of Mathematics, University of Maryland. Research partially supported by NSF DMS 880 1627 and an award from the Graduate Research Board of the University of Maryland
**Department of Mathematics, Boston University. Research partially supported by NSF DMS 8901913 and AFOSR-90-0017
†Sky Computers, Inc.

In [11], one of us (RL) used a geometric approach most similar in spirit to that of Carpenter. However, the technique of [11] was designed to give a somewhat stronger result: by constructing the desired solutions as the intersection of stable and unstable manifolds, one can obtain local uniqueness as well as existence. Furthermore the geometry exposed in the proof (specifically the "direction" of transversality) provides information that was crucial in proving the stability of these solutions (see Jones [12]).

There is one part of the proof in [11] which stands out as technically the most difficult; that is the part where the solutions are controlled by the slow flow. In recent work [13], the other two authors of the current paper have introduced a new technique that simplifies and considerably clarifies this part of the construction. This paper is then a new presentation of the proof of the travelling pulse for (1.1) as given in [11] but with this new method used to control the behavior near the slow manifold.

To exhibit the desired solution as an intersection of stable and unstable manifolds, it is necessary to follow the latter globally in space. As we show, one can factor the flow into a series of slow and fast parts; on a time scale in which the fast flow is order 1, the slow flow takes a time of order $1/\epsilon$. Nevertheless it turns out to be possible to follow an invariant manifold for that length of time. The key idea is that when the slow flow is dominant, the solution is near a "slow manifold" (to be defined below). The configuration of the tangent plane to the unstable manifold of the rest state changes as it passes near the slow manifold. Certain tangent directions are lost and converted into new ones which reflect the behavior on the slow manifold. We think of this as an exchange of information that occurs during the passage near the slow manifold, and thus we call the central result the "exchange lemma."

Our techniques for tracking an invariant manifold as it passes near a slow manifold are reminiscent of the "λ-lemma" (see [14]), which describes the behavior of an invariant manifold in the neighborhood of a hyperbolic critical point p. The latter says that if, at some time, the manifold has the dimension of $W^u(p)$ and is transverse to $W^s(p)$, then as $t \to \infty$ the manifold approaches $W^u(p)$. Our situation, however, is considerably more complex: instead of a critical point, we have a slow manifold of arbitrary dimension. The manifold approaches the product of the unstable directions with some center ones. The analogue of the "λ-lemma" would say that the manifold approaches the unstable directions, but this does not exhaust all the dimensions and the more subtle part lies in determining which center directions are picked out.

Our strategy is then to follow the tangent space to an invariant manifold through studying the induced flow on the space of exterior forms. The coordinates of a tangent vector can be viewed as 1-forms; analogously, a tangent plane has coordinates that are 2-forms. (Each is the area of the projection of the unit square in the tangent space onto a coordinate plane.) These evolve according to an easily computed equation and we control the tangent planes through estimates on their solutions.

Our goal in this paper is then twofold. First, we wish to make the geometric proof of [11] widely available. Secondly, we show how the techniques of [13] can be used in a basic example.

2. FitzHugh-Nagumo Equations. The travelling wave equations for the FHN system are given by (1.2). We shall show that these equations have an orbit that is homoclinic to $(0,0,0)$ when ϵ is sufficiently small. The orbit is constructed close to a certain "singular" solution, which is a union of trajectories for the $\epsilon = 0$ equation.

Setting $\epsilon = 0$ in (1.2), the equation for w is $w' = 0$. Hence w acts as a parameter in the other equations

$$\begin{aligned} u' &= v \\ v' &= \theta v - f(u) + w. \end{aligned} \tag{2.1}$$

We shall sometimes append to (2.1) one or the other of the auxiliary equations $w' = 0$ and $\theta' = 0$. In each case, which augmented equation of (1) we are considering will be made clear by the specific variables written.

For each θ and w, (2.1) has the rest point (u, v, w), with $v = 0$ and $w = f(u)$. For fixed θ there are two branches of the graph of $w = f(u)$ having negative slope. We call the associated curves of rest points S_L and S_R, where L and R denote the left and right hand branches respectively. See Figure 1. Aside from the points (u, v, w) at which $f'(u) = 0$, each such point is hyperbolic in its (u, v) plane, with θ and w fixed. Thus, each $p \in S_L \cup S_R$ has a well defined unstable and stable manifold $W^u(p)$ and $W^s(p)$.

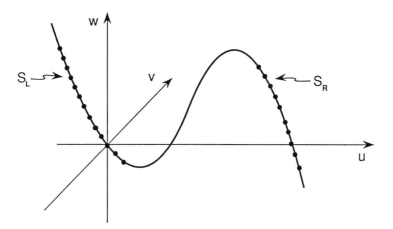

Figure 1. The rest points of 2.1. Of most interest are the two curves S_L and S_R on which the graph of $w = f(u)$ has negative slope.

For $w = 0$ and each θ, the rest point $(u, v) = (0, 0)$ in S_L is a rest point for the full system (1.2). It is well known that there is a value of θ, which we shall denote by

θ_*, for which there exists a connecting orbit in $w \equiv 0$ from $(u, v, w) = (0, 0, 0) \in S_L$ to $(u, v, w) = (1, 0, 0) \in S_R$. (See for instance Aronson and Weinberger [15].) The first piece of the singular solution is this connecting orbit.

By an argument similar to the one that determines θ_*, it can be shown that for $\theta = \theta_*$ there is a w_* for which (2.1) possesses a connecting orbit from $(u_-, 0, w_*) \in S_R$ to $(u_+, 0, w_*) \in S_L$, where u_- and u_+ satisfy $w_* = f(u)$. The second piece of the singular solution is the subset S_R with $\theta = \theta_*$ and $0 \le w \le w_*$. The third is the connecting orbit from $(u_-, 0, w_*)$ to $(u_+, 0, w_*)$. The last is the portion of S_L with $\theta = \theta_*$, traversed downward from $w = w_*$ to $w = 0$. See Figure 2. The singular solution is not smooth, since it has corners at the ends of each piece.

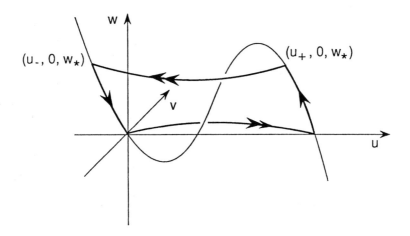

Figure 2. The singular orbit associated with $\theta = \theta_*$.

We can now state the theorem on the existence and uniqueness of the pulse solution to (1.2).

THEOREM: *If ϵ is sufficiently small, then for some θ within $0(\epsilon)$ of θ_*, (1.2) possesses an orbit which is homoclinic to the rest point (0,0,0) of (1.2). Moreover, this orbit lies within $0(\epsilon)$ of the singular solution S, and it is the unique such homoclinic orbit in such a neighborhood.*

The strategy of the proof is to display the homoclinic orbit as the transverse intersection of the unstable and stable manifolds of $(u, v, w) = (0, 0, 0)$ for some $\theta = \theta(\epsilon)$, $\epsilon \ne 0$ small. To do this, one shoots forward using the union of the unstable manifolds over a range of θ. Abusing notation, we shall use (1.2) to also denote the system with the equation $\theta' = 0$ appended. The union of unstable manifolds mentioned above is then a portion of the center-unstable manifold of (1.2). We denote this by $W^{cu} = \cup W^u(0, \theta)$, where 0 is the origin in (u, v, w) space and the union is over $\theta \in [\theta_* - \delta, \theta_* + \delta]$ for some appropriately small δ.

The difficulty of the proof is in following W^{cu} in order to show that, for some θ, $W^u(0, \theta)$ has a transverse intersection with $W^s(0, \theta)$. It is fairly straightforward to follow the manifold over the front, where each trajectory in W^{cu} remains close to the connecting trajectory of (2.1) from (0,0,0) to (1,0,0). The next and hardest step is to follow W^{cu} as it is carried close to the manifold S_R. This is where the exchange lemma is used. W^{cu} is then followed over the back. At the end of this third step, one is in a position to check the transversal intersection with the center stable manifold. The transversality calculations are carried out in (u, v, w, θ) space, and the locus of intersection determines the value of θ for the travelling wave. The transversality calculations then show that for that value of θ, $W^u(0, \theta)$ has a transverse intersection with $W^s(0, \theta)$.

The basic building blocks of the proof are the exchange lemma and two transversality lemmas involving only the behavior of the equations when $\epsilon = 0$. The latter results involve geometric objects built up out of the stable and unstable manifolds at $\epsilon = 0$ of the points of S_L and S_R. The first result is transversality along the front, as the wave speed θ varies: For each θ near θ_*, consider (2.1) with $w = 0$. The critical point $(u, v) = (0, 0)$ has a one dimensional unstable manifold. The center unstable manifold $W^{cu}(0, 0, \theta_*)$ of $(u, v, \theta) = (0, 0, \theta_*)$ is a two-dimensional manifold that is the union of the one-dimensional unstable manifolds. Similarly, at $w = 0$ the rest point $(u, v) = (1, 0)$ has a one dimensional stable manifold. The center stable manifold $W^{cs}(1, 0, \theta_*)$ of $(u, v, \theta) = (1, 0, \theta_*)$ is a two-dimensional manifold that is the union of the stable manifolds. These two-dimensional manifolds intersect along the connecting orbit at $\theta = \theta*$. The first transversality result is that this pair of two dimensional manifolds are transverse in (u, v, θ) space along the front.

LEMMA 1. $W^{cu}(0, 0, \theta_*)$ intersects $W^{cs}(1, 0, \theta_*)$ transversely in (u, v, θ) space, $w = 0$.

The other transversality result concerns the back, as θ is held fixed at θ_* and w is varied. As above, the rest points $(u_+, 0, w_*) \in S_R$ and $(u_-, 0, w_*) \in S_L$ have two-dimensional center unstable and center stable manifolds in (u, v, w) space that intersect along the connecting orbit in $w = w_*$.

LEMMA 2. $W^{cu}(u_-, 0, w_*)$ intersects $W^{cs}(u_+, 0, w_*)$ transversely in (u, v, w) space, $\theta = \theta_*$.

The proofs of lemmas 1 and 2 are well known. However, we take the opportunity in Section 4 to show how these can also be done very simply using differential forms.

We shall soon state the lemmas that will be used to describe the behavior of W^{cu} as it flows near S_R. We first need to define some objects associated with the $\epsilon = 0$ flow. Let $\overline{S}_R$ be a compact portion of S_R that includes $-\delta \leq w \leq w_* + \delta$ and let B denote a neighborhood in (u, v, w, θ) space of the set $\overline{S}_R \times I_\delta$, where $I_\delta = [\theta_* - \delta, \theta_* + \delta]$. Let $W^u(\overline{S}_R \times I_\delta)$ and $W^s(\overline{S}_R \times I_\delta)$ be the three-dimensional unions of unstable and stable manifolds of $\overline{S}_R \times I_\delta$ for the $\epsilon = 0$ equation (2.1). See Figure 3. Let $W^u(\overline{S}_R)$ be the two-dimensional restriction of $W^u(\overline{S}_R \times I_\delta)$ to $\theta = \theta_*$; we think of $W^u(\overline{S}_R)$ as a subset of (u, v, w, θ) space. (Note that $W^u(\overline{S}_R)$ includes what we previously called $W^{cu}(u_-, 0, w_*)$. We were then focusing on values of w near w_* and we are now interested in a larger portion of S_R.) The following two lemmas are direct consequences of the exchange lemma to be discussed in the next section. In the proof of the existence of the homoclinic orbit, they will be used to relate the $\epsilon \neq 0$ manifold W^{cu} that we are trying to follow to the singular object $W^u(\overline{S}_R)$ about which we have transversality information from Lemma 2.

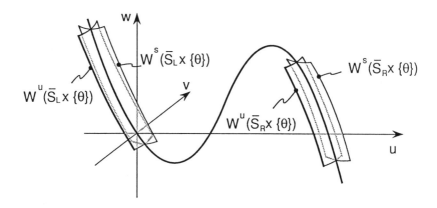

Figure 3. Slice in u, v, w, θ space of the stable and unstable manifolds of $\overline{S}_R$ and $\overline{S}_L$ for fixed θ. The slices are two-dimensional; $W^s(\overline{S}_R \times I_\delta)$ etc. are all three-dimensional in u, v, w, θ space.

Let Γ_ϵ be a smooth curve in ∂B, smoothly parametrized by ϵ, that is transverse to $W^s(\overline{S}_R \times I_\delta)$. See Figure 4. Assume the points $y_\epsilon \in \Gamma_\epsilon$ have w coordinates near $w = 0$. Denote by $z_\epsilon \in \partial B$ the point at which the trajectory through y_ϵ exits B. The first lemma says that the exit points lie close to the two-dimensional manifold $W^u(\overline{S}_R)$. Moreover, the "entrance curve" Γ_ϵ gets stretched under the flow into an "exit curve" that is C^0 close to $W^u(\overline{S}_R) \cap \partial B$. See Figure 4.

LEMMA 3. If $y_\epsilon \in \Gamma_\epsilon$ is sufficiently close to W^s then z_ϵ is within $0(\epsilon)$ of $W^u(\overline{S}_R)$. Moreover, for each point $p \in W^u(\overline{S}_R) \cap \partial B$ whose w coordinate is bounded away from 0 by a fixed amount, there exists a $y_\epsilon \in \Gamma_\epsilon$ whose associated z_ϵ is within $0(\epsilon)$ of p.

Let G_ϵ be the two-dimensional manifold that Γ_ϵ sweeps out under the flow of (1.2). It is an immediate corollary of Lemma 3 that for any $z_\epsilon \in G_\epsilon$ having w coordinates bounded away from 0, a neighborhood of z_ϵ in G_ϵ is C^0 close to $W^u(\overline{S}_R)$. To use transversality information from Lemma 2, it is necessary to have G_ϵ be C^1 close to $W^u(\overline{S}_R)$. That result is the following deeper fact, which is a consequence of the exchange lemma.

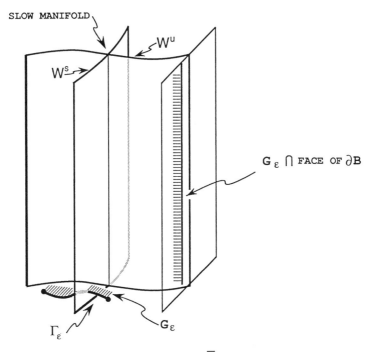

Figure 4. A curve Γ_ϵ transverse to $W^s(\overline{S}_R \times I_\delta)$ in u, v, w, θ space. The θ direction is suppressed in this figure. The θ component varies strongly along Γ_ϵ. However, it is a consequence of Lemma 4 that the image curve on $G_\epsilon \cap 2B$ has almost no dependence on θ.

LEMMA 4. *If y_ϵ is sufficiently close to W^s, then there is a neighborhood $N(z_\epsilon)$ so that $G_\epsilon \cap N(z_\epsilon)$ is C^1 close to $W^u(\overline{S}_R)$.*

Note that the entrance curve Γ_ϵ may have points with a range of values of θ. Nevertheless, the C^1 closeness of G_ϵ to $W^u(\overline{S}_R)$ at the exit set implies that the exit curve has a tangent vector whose θ component is almost zero, i.e. the θ information has been washed away.

The proof of the theorem can now be given. As mentioned above, we follow around W^{cu} for ϵ small. If ϵ is small enough, W^{cu} lies close to the set $w = 0$ until the trajectory for $\theta = \theta_*$ enters a neighborhood of S_R. It therefore lies close to $W^{cu}(0, 0, \theta_*)$, the $\epsilon = 0$ singular object in $w = 0$. Lemma 1 states that $W^{cu}(0, 0, \theta_*)$ is transverse to the two-dimensional $W^{cs}(1, 0, \theta_*)$ in $w = 0$. It is an easy corollary that $W^{cu}(0, 0, \theta_*)$ is transverse to the three dimensional $W^s(\overline{S}_R \times I_\delta)$ in (u, v, w, θ) space (see Figure 5). Thus we may conclude that W^{cu} is transverse to $W^s(\overline{S}_R \times I_\delta)$ at a point in $\Gamma_\epsilon \equiv W^{cu} \cap \partial B$.

We can now apply Lemmas 3 and 4. It follows from Lemma 3 that for any value w near w_*, there are points entering B on Γ_ϵ that exit at height w near the singular object $W^{cu}(u_-, 0, w_*)$. Furthermore, by Lemma 4, at any such exit point z_ϵ, the tangent plane to W^{cu} is close to that of the singular object $W^u(\overline{S}_R)$ or,

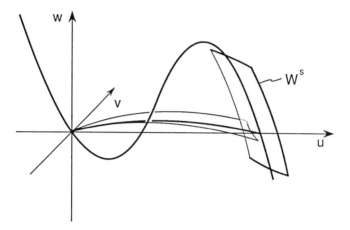

Figure 5. A schematic drawing of the intersection of the two-dimensional manifold $W^{cu}(0,0,\theta_*) = \cup_{\theta \in I_\delta} W^u(0,0,\theta)$ with the three dimensional manifold $W^s(\overline{S}_R \times I_\delta)$. The θ direction has been suppressed by projecting onto (u,v,w) space. For $\theta \approx \theta_*$, the projection is approximately $W^s(\overline{S}_R \times \{\theta_*\})$.

equivalently, to that of $W^{cu}(u_-, 0, w_*)$.

We may now use Lemma 2 to follow W^{cu} across the back. By Lemma 2, $W^{cu}(u_-, 0, w_*)$ has a transverse intersection with $W^{cs}(u_+, 0, w_*)$ in (u,v,θ) space; it follows that $W^{cu}(u_-, 0, w_*)$ has a transverse intersection in (u,v,w,θ) space with the three-dimensional manifold $W^s(\overline{S}_L \times I_\delta)$. (The latter is defined analogously to $W^s(\overline{S}_R \times I_\delta)$ to be the singular object that is the union over $-\delta \leq w \leq w_* + \delta$, $\theta \in I_\delta$, of the stable manifolds associated to points on S_L at $\epsilon = 0$.) The C^1- closeness of W^{cu} to $W^{cu}(u_-, 0, w_*)$ implies that it too has a transverse intersection with $W^s(\overline{S}_R \times I_\delta)$.

The final observations concern the center stable manifold W^{cs} of the point $(u,v,w) = (0,0,0)$ for (1.2). If $\epsilon \neq 0$, for each $\theta \in I_\delta$, the center stable manifold of this point is two-dimensional (one fast direction and one slow), so W^{cs} is a three-dimensional manifold in (u,v,w,θ) space. For $\epsilon \neq 0$ small, W^{cs} lies close to $W^s(\overline{S}_L \times I_\delta)$, at least in a neighborhood of $\overline{S}_L$. Thus we may conclude the transversality of W^{cu} and W^{cs}. Since the transverse intersection of a two-dimensional manifold and a three-dimensional manifold in four-space is a one-dimensional curve, this is the trajectory we seek. θ is a constant along trajectories, so the intersection defines the wave speed θ of the arc.

3. The Exchange Lemma. We shall now sketch ideas behind the main technical tool of the proof, the exchange lemma. Although this lemma has been proved in arbitrary dimensions [13], we discuss it here in the lowest interesting dimensions, which is all that is needed for the Fitzhugh-Nagumo equations.

The exchange lemma uses special coordinate systems adapted to a neighborhood of S_L or S_R. At $\epsilon = 0$, for fixed θ, it is easy to use the hyperbolic structure of the rest points to get a coordinate system (a,b) in which the equations have the form

$$
\begin{aligned}
a' &= \lambda(a, b, \mathbf{w})a \\
b' &= \mu(a, b, \mathbf{w})b \\
\mathbf{w}' &= 0
\end{aligned}
$$

(3.1)

where $a, b, \epsilon R^1$, $\lambda > 0$ and $\mu < 0$. $\mathbf{w} \epsilon R^2$ and represents the (w, θ) coordinates. For any $\delta > 0$ sufficiently small, there are λ_0, μ_0 such that

(3.2) $$\lambda(a, b, \mathbf{w}) \geq \lambda_0 > 0, \quad \mu(a, b, \mathbf{w}) \leq \mu_0 < 0$$

for all $B \equiv \{a, b, \mathbf{w} : |a| < \delta, |b| < \delta, \mathbf{w} \text{ in a given compact region}\}$. For $\epsilon \neq 0$, there are analogues $S_{L,\epsilon}$ and $S_{R,\epsilon}$ of the manifolds S_L and S_R. These are constructed (nonuniquely) as center manifolds of (1.2). On them, the vector fields have size $0(\epsilon)$, so $S_{L,\epsilon}$ and $S_{R,\epsilon}$ are known as "slow manifolds." It was shown by Fenichel [16] that in the neighborhood B of S_L or S_R, there is an analogue of the coordinate system (3.1) adapted to the hyperbolic structure that persists, with the stable and unstable coordinates vanishing on $S_{L,\epsilon}$ or $S_{R,\epsilon}$. In this coordinate system, which we shall call "Fenichel coordinates," the equations near $S_{L,\epsilon}$ or $S_{R,\epsilon}$ take the form

$$
\begin{aligned}
a' &= \lambda(a, b, \mathbf{w}, \epsilon)a \\
b' &= \mu(a, b, \mathbf{w}, \epsilon)b \\
\mathbf{w}' &= \epsilon g(a, b, \mathbf{w}, \epsilon)
\end{aligned}
$$

(3.3)

We may assume that $g = (g_1, g_2)$ satisfies

(3.4)
$$
\begin{aligned}
g_1(0, 0, w_1, w_2, 0) &\geq c > 0 \\
g_2(0, 0, w_1, w_2, 0) &= 0
\end{aligned}
$$

That is, the $\mathbf{w}$ coordinates are chosen so that w_2 vanishes along the limiting trajectory of the slow flow as $\epsilon = 0$. We also assume that the box B has the form $w_1^- < w_1 < w_1^+; |w_2|, |a|, |b| < \delta$, and that at $\epsilon = 0$, the critical points $a = 0$, $b = 0$, $w_1^- < w_1 < w_1^+, |w_2| < \delta$ form a normally hyperbolic manifold. These coordinates, also used in [11], simplify the calculations we shall do.

Let M be some two-dimensional invariant manifold that has an intersection with the manifold $\{b = \delta\}$ at some point q. As we shall show, the trajectory through q, which lies in M by invariance, enters the interior of B. If the a coordinate of q is in an appropriate range, the trajectory through q remains in B for a time that is $0(1/\epsilon)$ and leaves at a point $\overline{q}$ in a face of the form $|a| = \delta$. We shall be concerned with the relationship between the tangent plane to M at q and that of the tangent plane to M at $\overline{q}$.

The exchange lemma says that under appropriate restrictions on the tangent plane at q, including transversality with respect to the space $\{a = 0\}$, the tangent

The exchange lemma uses special coordinate systems adapted to a neighborhood of S_L or S_R. At $\epsilon = 0$, for fixed θ, it is easy to use the hyperbolic structure of the rest points to get a coordinate system (a,b) in which the equations have the form

(3.1)
$$a' = \lambda(a, b, \mathbf{w})a$$
$$b' = \mu(a, b, \mathbf{w})b$$
$$\mathbf{w}' = 0$$

where $a, b, \epsilon R^1$, $\lambda > 0$ and $\mu < 0$. $\mathbf{w} \epsilon R^2$ and represents the (w, θ) coordinates. For any $\delta > 0$ sufficiently small, there are λ_0, μ_0 such that

(3.2)
$$\lambda(a, b, \mathbf{w}) \geq \lambda_0 > 0, \quad \mu(a, b, \mathbf{w}) \leq \mu_0 < 0$$

for all $B \equiv \{a, b, \mathbf{w} : |a| < \delta, |b| < \delta, \mathbf{w} \text{ in a given compact region}\}$. For $\epsilon \neq 0$, there are analogues $S_{L,\epsilon}$ and $S_{R,\epsilon}$ of the manifolds S_L and S_R. These are constructed (nonuniquely) as center manifolds of (1.2). On them, the vector fields have size $0(\epsilon)$, so $S_{L,\epsilon}$ and $S_{R,\epsilon}$ are known as "slow manifolds." It was shown by Fenichel [16] that in the neighborhood B of S_L or S_R, there is an analogue of the coordinate system (3.1) adapted to the hyperbolic structure that persists, with the stable and unstable coordinates vanishing on $S_{L,\epsilon}$ or $S_{R,\epsilon}$. In this coordinate system, which we shall call "Fenichel coordinates," the equations near $S_{L,\epsilon}$ or $S_{R,\epsilon}$ take the form

(3.3)
$$a' = \lambda(a, b, \mathbf{w}, \epsilon)a$$
$$b' = \mu(a, b, \mathbf{w}, \epsilon)b$$
$$\mathbf{w}' = \epsilon g(a, b, \mathbf{w}, \epsilon)$$

We may assume that $g = (g_1, g_2)$ satisfies

(3.4)
$$g_1(0, 0, w_1, w_2, 0) \geq c > 0$$
$$g_2(0, 0, w_1, w_2, 0) = 0$$

That is, the $\mathbf{w}$ coordinates are chosen so that w_2 vanishes along the limiting trajectory of the slow flow as $\epsilon = 0$. We also assume that the box B has the form $w_1^- < w_1 < w_1^+; |w_2|, |a|, |b| < \delta$, and that at $\epsilon = 0$, the critical points $a = 0$, $b = 0$, $w_1^- < w_1 < w_1^+, |w_2| < \delta$ form a normally hyperbolic manifold. These coordinates, also used in [11], simplify the calculations we shall do.

Let M be some two-dimensional invariant manifold that has an intersection with the manifold $\{b = \delta\}$ at some point q. As we shall show, the trajectory through q, which lies in M by invariance, enters the interior of B. If the a coordinate of q is in an appropriate range, the trajectory through q remains in B for a time that is $0(1/\epsilon)$ and leaves at a point $\overline{q}$ in a face of the form $|a| = \delta$. We shall be concerned with the relationship between the tangent plane to M at q and that of the tangent plane to M at $\overline{q}$.

The exchange lemma says that under appropriate restrictions on the tangent plane at q, including transversality with respect to the space $\{a = 0\}$, the tangent plane at $\overline{q}$ is C^1 close to the plane spanned by the a and w_1 axes. Let $\gamma = M \cap \{b = \delta\}$ be an open arc, and $\overline{\gamma}$ a subset of γ such that points on $\overline{\gamma}$ have trajectories that exit B through $\{a = \delta\}$. The exchange lemma implies that the map $W: \overline{\gamma} \to \{a = \delta\}$ is smooth, and that the tangent vector to the image $W(\overline{\gamma})$ at a point $\overline{q}$ is C^1 close to the w_1 direction. In particular, information about transversality with respect to $\{a = 0\}$ at the entry to B is exchanged for information about transversality with respect to w_1. In some applications, including FitzHugh-Nagumo, knowledge about transversality with respect to $\{a = \delta\}$ at the entry comes from information about how the limiting unstable spaces $W^u(p)$ vary as the other slow variable w_2 is changed. Thus information about behavior as w_2 is varied is exchanged for information about the behavior of w_1.

It is not hard to establish that, while a trajectory is in B, its b coordinate decreases exponentially and its a coordinate increases exponentially [13]. Also, the time from entrance to exit is $0(1/\epsilon)$ for any point whose w_1 coordinate increases by $0(1)$. This quickly implies that if V is a sufficiently small neighborhood of q in M_ϵ, then the image of V at the exit of B is C^0 close to $\{w_2 = 0, b = 0\}$.

It is more subtle to establish that the image is C^1 close to the above two-dimensional plane, since it is hard to see the behavior of planes by following individual vectors. The difficulty is that every trajectory, when exiting from B, does so approximately tangent to the expanding subspace parameterized by the a variable. Thus, the study of individual tangent vectors does not pick up the part of the tangent plane to M in the center direction. To do this, one must directly follow how the flow induces a vector field on tangent planes. This flow is computed from the variational flow on $\mathbf{y} \equiv (a, b, w_1, w_2)$, which is

$$
\begin{aligned}
\delta a' &= \lambda(\mathbf{y}, \epsilon)\delta a + a(\nabla\lambda \cdot \delta\mathbf{y}) \\
\delta b' &= \mu(\mathbf{y}, \epsilon)\delta b + b(\nabla\mu \cdot \delta\mathbf{y}) \\
\delta w_1' &= \epsilon\nabla g_1 \cdot \delta\mathbf{y} \\
\delta w_2' &= \epsilon\nabla g_2 \cdot \delta\mathbf{y}.
\end{aligned}
$$

(3.5)

We may think of each of the above coordinates $\delta a, \delta b, \delta w_1, \delta w_2$ as one-forms acting on the vector fields, and the variational equations are evolution equations for the one-forms. A plane is represented by a two-vector, which is an element of the second exterior power $\Lambda^2 \mathbf{R}^4 \cong \mathbf{R}^6$. The coordinates of this vector in $\mathbf{R}^6$ can be thought of as two-forms. To follow planes, we derive evolution equations for the two-forms. The space of two-forms is spanned by the six two-forms $P_{\alpha\beta} \equiv \delta\alpha \wedge \delta\beta$, where $\alpha, \beta = a, b, w_1, w_2$. (There are only six independent such ones since $\delta\alpha \wedge \delta\alpha = 0$ and $\delta\alpha \wedge \delta\beta = -\delta\beta \wedge \delta\alpha$.) The evolution equations for the two-forms can be computed from (3.5) using the product rule:

$$P'_{aw_1} = \lambda P_{aw_1} + aR_1 + \epsilon R_2$$
$$P'_{aw_2} = \lambda P_{aw_2} + aR_3 + \epsilon R_4$$
(3.6)
$$P'_{ab} = [\lambda + \mu]P_{ab} + aR_5 + bR_6$$
$$P'_{bw_1} = \mu P_{bw_1} + bR_7 + \epsilon R_8$$
$$P'_{bw_2} = \mu P_{bw_2} + bR_9 + \epsilon R_{10}$$
$$P'_{w_1 w_2} = \epsilon(R_{11} + R_{12}),$$

where

$$R_1 = (\nabla\lambda \cdot \delta\mathbf{y}) \wedge \delta w_1 \qquad R_7 = (\nabla\mu \cdot \delta\mathbf{y}) \wedge \delta w_1$$
$$R_2 = \delta a \wedge (\nabla g_1 \cdot \delta\mathbf{y}) \qquad R_8 = \delta b \wedge (\nabla g_1 \cdot \delta\mathbf{y})$$
$$R_3 = (\nabla\lambda \cdot \delta\mathbf{y}) \wedge \delta w_2 \qquad R_9 = (\nabla\mu \cdot \delta\mathbf{y}) \wedge \delta w_2$$
$$R_4 = \delta a \wedge (\nabla g_2 \cdot \delta\mathbf{y}) \qquad R_{10} = \delta b \wedge (\nabla g_2 \cdot \delta\mathbf{y})$$
$$R_5 = (\nabla\lambda \cdot \delta\mathbf{y}) \wedge \delta b \qquad R_{11} = \delta w_1 \wedge (\nabla g_1 \cdot \delta\mathbf{y})$$
$$R_6 = \delta a \wedge (\nabla\mu \cdot \delta\mathbf{y}) \qquad R_{12} = (\nabla g_2 \cdot \delta\mathbf{y}) \wedge \delta w_2$$

It is useful to do the calculations in $\mathbf{P} \wedge^2 \mathbf{R}^4$, projective space of two-forms in $\mathbf{R}^4$. The reason is that the values of the two-forms acting on a plane are determined only up to normalization. Also, by (3.6), values of some of the two-forms grow exponentially in time. We can use the fastest growing two-form to normalize. In a region in which $P_{a,w_1} \neq 0$, $\{P_{\alpha\beta}/P_{aw_1}\}$ form local coordinates, with $\alpha, \beta = a, b, w_1, w_2$ and $(\alpha, \beta) \neq (a, w_1)$ or (w_1, a). Let $\hat{P}_{\alpha\beta} = P_{\alpha\beta}/P_{aw_1}$ and $\hat{R}_i = R_i/P_{aw_1}$. The $\{\hat{R}_i\}$ can be written in terms of the $\{\hat{P}_{\alpha\beta}\}$ and a nonhomogeneous term (time dependent). In these coordinates, the plane spanned by the directions a and w_1 is specified by $\hat{P}_{\alpha\beta} = 0$. A plane with $\hat{P}_{\alpha\beta} = 0(\epsilon)$ is C^1 close to this plane. We shall sketch the proof of:

EXCHANGE LEMMA: Let M be a two-dimensional invariant manifold transverse to $\{b = \delta\}$. Assume that $\Gamma = M \cap \{b = \delta\}$ is transverse to $\{a = 0\}$. Let q in Γ be a point whose trajectory exits from $\{a = \delta\}$ after a time τ that is $0(1/\epsilon)$. Then at the exit point $\overline{q}, \hat{P}_{\alpha\beta}$ is $0(\epsilon)$ for all α, β.

Sketch: The evolution equations for the $\{\hat{P}_{\alpha\beta}\}$ are

$$\hat{P}'_{aw_2} = \hat{P}_{aw_2}(-Q) + a\hat{R}_3 + \epsilon\hat{R}_4$$
$$\hat{P}'_{ab} = (\mu - Q)\hat{P}_{ab} + a\hat{R}_5 + b\hat{R}_6$$
(3.7)
$$\hat{P}'_{bw_1} = (\mu - \lambda - Q)\hat{P}_{bw_1} + b\hat{R}_7 + \epsilon\hat{R}_8$$
$$\hat{P}'_{bw_2} = (\mu - \lambda - Q)\hat{P}_{bw_2} + b\hat{R}_9 + \epsilon\hat{R}_{10}$$
$$\hat{P}'_{w_1 w_2} = (-\lambda - Q)P_{w_1 w_2} + \epsilon(\hat{R}_{11} + \hat{R}_{12}),$$

where $Q = a\hat{R}_1 + \epsilon\hat{R}_2$.

One first shows that all $\hat{P}_{\alpha\beta}$ decay exponentially to size $0(\epsilon)$, provided that $\hat{P}_{aw_2}$ satisfies some a-priori bound. This follows (with some work) from the fact that the lower R.H. 4×4 matrix of (3.7) is almost diagonal, with negative diagonal elements. The less intuitive part of the proof is to show that $\hat{P}_{aw_2}$ must have started small, and stays small. (It therefore satisfies the a-priori estimates.)

Consider the initial conditions for the $\{\hat{P}_{\alpha\beta}\}$, including P_{aw_2}. The tangent plane to M at q is spanned by the vector $\mathbf{y}' = (a', b', w_1', w_2')$ at q and the tangent vector $d\Gamma$ to $M \cap \{b = \delta\}$. In the first vector, $(1/\epsilon)dw_1/d\tau \geq c > 0$. $db/d\tau$ is $0(1)$. By hypothesis, the trajectory through q remains inside B for a time that is $0(1/\epsilon)$. Since a increases exponentially, at q the value of a is exponentially small; hence so is $da/d\tau$. By the transversality hypothesis on Γ, the first coordinate of $d\Gamma$ is bounded away from zero independent of ϵ. The second is zero by definition of Γ. The value of the two-form $P_{\alpha\beta}$ is computed up to a normalization factor by taking the 2×2 determinant of the α, β components of the above two vectors. (See Section 4 for more details.) From that, we obtain that P_{aw_1} is $0(\epsilon)$, with its quotient by ϵ bounded away from zero. P_{aw_2} is exponentially small, and the other two-forms, evaluated on $T_q M$, have $0(1)$ bounds. It follows that $\hat{P}_{aw_2}$ is also exponentially small in ϵ.

It remains to show that $\hat{P}_{aw_2}$ stays $o(1)$. This uses the first equation of (3.7) and information about R_3 and R_4 to control the nonhomogeneous term $a\hat{R}_3 + \epsilon\hat{R}_4$. The technical lemmas and estimates are in [13].

4. Proofs of Transversality.

The proofs of lemmas 1 and 2 can also be achieved through the use of forms. Consider first Lemma 1. We abbreviate $W^{cu}(0, 0, \theta_*)$ and $W^{cs}(1, 0, \theta_*)$ by W^{cu} and W^{cs} respectively, warning the reader not to confuse these with the $\epsilon \neq 0$ objects so denoted in the previous sections. We need to show that the tangent spaces to W^{cu} and W^{cs} intersect transversely in (u, v, θ) space. We calculate the 2-forms which give us the coordinates associated with these tangent planes; it suffices to show that the vectors made out of these coordinates are linearly independent.

The variational equations are

$$(4.1) \quad \begin{aligned} \delta u' &= \delta v \\ \delta v' &= \theta\delta v - f'(u)\delta u + v\delta\theta \\ \delta\theta' &= 0. \end{aligned}$$

The equations for the 2-forms are calculated as in the previous section with $P_{uv} = \delta u \wedge \delta v$ etc. The relevant equation is the one for P_{uv}:

$$(4.2) \quad P_{uv}' = \theta P_{uv} + v P_{u\theta}.$$

We are interested in the P_{uv} and $P_{u\theta}$ associated with the tangent planes to W^{cu} and W^{cs}, which we shall denote $P_{uv}^{\pm}$ and $P_{u\theta}^{\pm}$ (+ for unstable and − for stable). We shall show that $P_{u\theta}^+$ and P_{uv}^+ have the same sign, whereas $P_{u\theta}^-$ and P_{uv}^- have opposite signs. This will imply that the vectors $(P_{uv}, P_{u\theta}, P_{v\theta})$ for the tangent planes to W^{cs} and W^{cu} are linearly independent and hence that the planes are transverse. We explicitly calculate $P_{u\theta}^{\pm}$ and then obtain the sign of $P_{uv}^{\pm}$ from (4.2).

We must first be explicit about how to associate a number to a 2-form acting on a plane. The plane is specified by giving an orthonormal pair of tangent vectors whose span determines an oriented unit square in the plane. Consider a 2-form $\delta\alpha_1 \wedge \delta\alpha_2$, where the α_i vary over u, v, θ. The value of $\delta\alpha_1 \wedge \delta\alpha_2$ on the plane is given, up to sign, by the area of the projection of the above square on the α_1, α_2 plane along α_3, where $\alpha_3 \neq \alpha_1, \alpha_2$. The sign is positive if the orientation of the projected square, induced by the projection, agrees with the orientation of the α_1, α_2 plane induced from the ordering of α_1, α_2.

Now we return to W^{cs} and W^{cu}. Both of these planes have the vector field $(v, \theta v - f(u), 0)$ as a tangent vector. Let $(\delta u^\pm, \delta v^\pm, 1)$ be another tangent vector for W^{cu} and W^{cs} respectively ($+$ for unstable and $-$ for stable). Note that we can take $\delta\theta = 1$; since $\delta\theta' = 0$, this guarantees that the new vector is not a multiple of the vector field. The above vectors are not orthonormal, but they can be made orthonormal by normalization and a Gramm-Schmidt process. The quantities P_{uv}, etc., for each plane are thus equal, up to a normalization factor, to the 2×2 subdeterminants of the 2×3 matrix whose rows are the above linearly independent vectors in (u, v, θ) space spanning the plane. The normalization factor $N = N(u, v, \delta u^\pm, \delta v^\pm)$ is positive and comes from the orthonormalization procedure. Thus

$$P_{u\theta}^\pm = N \begin{vmatrix} v & 0 \\ \delta u^\pm & 1 \end{vmatrix} = Nv$$

Since $v > 0$ along the front, for the tangent planes to both W^{cu} and W^{cs}, we then have that $P_{u\theta}^\pm > 0$. Moreover (4.2) now reads

(4.3) $$P_{uv}^{\pm\prime} = \theta P_{uv}^\pm + v^2 N$$

W^{cu} and W^{cs} each contain a line of critical points whose tangent vector is in the θ direction. For any plane containing such a line, the 2-form P_{uv} vanishes. It follows that $P_{uv}^+ \to 0$ as $\xi \to -\infty$ and $P_{uv}^- \to 0$ as $\xi \to +\infty$. Equation (4.3) then implies that for W^{cu}, $P_{uv}^+ > 0$ and for W^{cs}, $P_{uv}^- < 0$. This proves that these vectors associated with the respective tangent planes are linearly independent, as desired. $\square$

The proof of Lemma 2 is similar. The equations are now

$$\delta u' = \delta v$$
$$\delta v' = \theta_* \delta v - f'(u)\delta u + \delta w$$
$$\delta w' = 0$$

The relevant equation on 2-forms is

$$P_{uv}' = \theta_* P_{uv} + P_{uw}.$$

Also, $P_{uw} = \overline{N}v$, where $\overline{N}$ is the analogous normalization factor. Since v is of one sign ($v < 0$) along the back, a similar argument holds. $\square$

Acknowledgment.

The authors are grateful to J. Alexander for a number of discussions that clarified the use of forms. We thank the Institute for Mathematics and its Applications for its hospitality while this paper was shaped.

REFERENCES

[1] CONLEY, C., *On travelling wave solutions of nonlinear diffusion equations in dynamical systems, theory and applications*, in Springer Lecture Notes in Physics.

[2] CARPENTER, G., *A geometric approach to singular perturbation problems with applications to nerve impulse equations*, JDE, 23 (1977), 335–367.

[3] HASTINGS, S., *On travelling wave solutions of the Hodgkin-Huxley equations*, Arch. Rat. Mech. Anal., 60 (1976), 229–257.

[4] GARDNER, R.H. AND SMOLLER, J., *Travelling wave solutions of predator-prey systems with singularly perturbed diffusion*, JDE, 47 (1983), 133–161.

[5] FIFE, P.C., *Boundary and interior transition layer phenomena for pairs of second order differential equations*, J. Math. Anal. Appl., 54 (1976), 497–521.

[6] FUJII, H., NISHIURA, Y., AND HOSONO, Y., *On the structure of multiple existence of stable stationary solutions in systems of reaction-diffusion equations*, Studies in Mathematics and its Applications, 18 (1986), 157–219.

[7] HALE, J. AND SAKAMOTO, K., *Existence and stability of transition layers*, Japan J. Appl. Math, 5 (1988), 367–405.

[8] SAKAMOTO, K., *Invariant manifolds in singular perturbation problems for ordinary differential equations*, preprint.

[9] CHANG, K.W. AND HOWES, F.A., *Nonlinear Singular Perturbation Phenomena: Theory and Applications*, Springer Verlag, N.Y. 1984.

[10] DIENER, F. AND REEB, G., *Analyse Non Standard*, Hermann, Editeurs des Sciences et des Arts, PARIS, 1989.

[11] LANGER, R., *Existence and uniqueness of pulse solutions to the Fitzhugh-Nagumo equations*, Ph.D. Thesis, Northeastern University (1980).

[12] JONES, C.K.R.T., *Stability of the travelling wave solutions of the Fitzhugh-Nagumo system*, Trans. AMS, 286, #2 (1984), 431–469.

[13] JONES, C.K.R.T. AND KOPELL, N., *Tracking invariant manifolds of singularly perturbed systems using differential forms*, in preparation.

[14] PALIS, J. AND DE MELO, W., *Geometric Theory of Dynamical Systems*, Springer-Verlag, N.Y. 1980.

[15] ARONSON, D. AND WEINBERGER, H., *Nonlinear diffusions in population genetics, combustion and nerve propagation*, in Partial Differential Equations and Related Topics, J. Goldstein, ed., Lecture Notes in Math #446, Springer Verlag 1975.

[16] FENICHEL, N., *Geometric singular perturbation theory for ordinary differential equations*, J. Diff. Equa., 31 (1979), pp. 53–98.

KINETIC POLYNOMIAL:
A NEW CONCEPT OF CHEMICAL KINETICS

MARK Z. LAZMAN* AND GREGORY S. YABLONSKII**

Abstract. A system of quasi-steady-state equations for a single pathway mechanism of a catalytic reaction can always be reduced to a polynomial in terms of the steady state reaction rate, a kinetic polynomial. The coefficients of this polynomial are polynomials in the parameters of the elementary reaction rates. The form of the lowest coefficient of the polynomial ensures the thermodynamic validity of this form of representation of quasi-steady-state equations. The properties of the kinetic polynomial are discussed in connection with such concepts of chemical kinetics as "molecularity", "stoichiometric number".

Possible applications of this form are: asymptotic analysis of steady-state kinetic models as well as analysis of steady-state multiplicity; description of the steady-state dependences of the reaction rate, determining relations between kinetic constants when solving the inverse problem.

On the basis of kinetic polynomial explicit equations for the steady-state rate in case when one of the steps is rate-limiting, and in the neighbourhood of equilibrium have been derived.

Algorithm of computation of the kinetic polynomial and its realisation on the basis of computer algebra are described.

Key words. steady-state, catalytic reaction.

AMS(MOS) subject classifications.

1. Introduction. The general description of catalytic reaction is given in the theory of steady reaction (Horiuti-Temkin [1,2]). The mass (surface)-action-law (MAL) is assumed for the rate of elementary reaction in this theory. According to the steady reaction theory the system of quasi-steady-state equations is of the form

$$(1.1) \qquad \sum_{p=1}^{P} \nu_s^{(p)} w^{(p)} = u_{(s)}, \quad s = 1, \ldots, S; \quad p = 1, \ldots, P,$$

where $u_{(s)} = u_s - u_{-s}, u_s, u_{-s}$ are the rates of s-th stage and its forward and back reactions; $\nu_s^{(p)}$ is the stoichiometric number of the sth step for pth independent route[1]; $w^{(p)}$ is the rate along the pth route; S is the number of steps; $P = S - J$ is the number of reaction routes (J is the number of independent intermediates).

An important problem concerned with the steady reaction theory was the dependence between the kinetic equations of forward and back reactions or more generally the connection between thermodynamics and kinetics for complex reactions. The case with a rate-determining step has been studied by Boreskov [4] and, independently, by Horiuti [1,3]. They have established the dependence of the

*USSR Research and Design Institute of Oil Refining and Petrochemical Industry (VNIP-INEFT), Moscow 113095, Dimitrova 33/13, USSR.

**Tuva Complex Department of Siberian Branch USSR Academy of Sciences, Kyzyl 667000, Lenina 30, USSR.

[1] This concept was introduced by Horiuti [3]. The set of numbers $\nu_s^{(p)}$ provides that adding up the steps of complex reaction multiplied by respective $\nu_s^{(p)}$ results in a net chemical equation free from intermediates.

steady-state reaction rate w on the rate of forward reaction w_+ and thermodynamic characteristics of the overall reaction:

$$(1.2) \qquad w = w_+(1 - \exp(\Delta G/(\nu_L RT))) = w_+(1 - (f_-(\mathbf{c})/(K_{eq}f_+(\mathbf{c})))^M).$$

Here ΔG is free energy variations; K_{eq} is the equilibrium constant of the overall reaction; $f_+(\mathbf{c})$ and $f_-(\mathbf{c})$ are the products of concentrations written according to the MAL for net stoichiometric equation both forward and backward; $\mathbf{c}$ is the reactant concentration vector; M is the reaction molecularity and ν_L is the stoichiometric number of the rate-determining step $(M = \nu_L^{-1})$.

Near equilibrium the rate of a simple reaction is linearly dependent on the free energy variations [5]. Using this fact Nacamura [6] has obtained for a single-route reaction the following equation

$$(1.3) \qquad -\Delta G/(RT) = \left(\sum_{s=1}^{S} \nu_s^2/u_{seq}\right) w,$$

where ν_s and u_{seq} are the stoichiometric number and the equilibrium rate of exchange for the sth step.

Formulae of types (1.2) and (1.3) are of importance in the kinetics of chemical reactions [7]. They were used to study reaction mechanisms for NH_3 synthesis, SO_2 oxidation, etc. [8]. In many publications the possibility has been discussed for the extension of their applicability [9-12].

Further development of steady reaction theory was based on the graph theory that had been applied initially in enzyme kinetics [13]. A structured form of steady-state kinetic equations based on graph representation of catalytic reactions with linear mechanism[2] was found and investigated in [14]. This form enables writing explicitly a steady-state kinetic equation on the basis of a detailed mechanism. However, the main advantage of such forms is the possibility of obtaining physico-chemically meaningful corollaries rather than their compactness.

For instance, for a linear single pathway mechanism the steady-state reaction rate w is represented the following equation [14]:

$$(1.4) \qquad w = \left(\prod_{s=1}^{S} b_s - \prod_{s=1}^{S} b_{-s}\right) / \sum_{x} D_x,$$

where b_s and b_{-s} are the weights of the forward and back reactions, i.e. their rates at unit concentrations of the intermediates[3], and D_x is the weight of the framework (see [14]), i.e. the product of the weights of the component arcs.

---

[2] Only one intermediate reacts in each elementary step of these reaction mechanism, for example,

$$1)A + Z_1 \rightleftharpoons Z_2, \quad 2)Z_2 \rightleftharpoons Z_3, \quad 3)Z_3 \rightleftharpoons B + Z_1$$

(Z_1, Z_2, Z_3 are intermediates, A is reactant, and B is product).

[3] for mechanism in previous footnote the weights are: $b_1 = k_1 c_A, b_{-1} = k_{-1}, \ldots, b_3 = k_3, b_{-3} = k_{-3}c_B$; where k_i are the rate constants of corresponding steps.

The numerator in (1.4) corresponds to the kinetic equation of the overall reaction assuming that it is a step for which the MAL is valid [14]. One can transform it to the form $k_+ f_+(\mathbf{c}) - k_- f_-(\mathbf{c})$, thus the equation (1.4) corresponds the equation (1.2) at $M = 1$.

From Eq. (1.4) it follows that at $w = 0$ (equilibrium)

$$(1.5) \qquad \prod_{s=1}^{S} b_s = \prod_{s=1}^{S} b_{-s} = k_+ f_+(\mathbf{c}) - k_- f_-(\mathbf{c}) \text{ or } K_{eq} = f_-(\mathbf{c})/f_+(\mathbf{c}).$$

Eq. (1.5) is valid obviously for the kinetic equations (1.2) and (1.3).

We'll consider in this paper the general situation of nonlinear reaction mechanism. The first problem here is the form of kinetic equation. The main reason that there were no structured forms like (1.4) for nonlinear mechanisms in the literature has apparently been the impossibility of solving explicitly a system of quasi-steady-state nonlinear equations (1.1). However, it is always possible to apply to this system of equations a method of elimination of variables and reduce it to a polynomial in one variable, e.g. a polynomial in terms of the steady-state reaction rate w. The polynomial coefficients are, in turn, polynomials in terms of the reaction rate constants (k_s and k_{-s}) and reactant concentrations ($\mathbf{c}$). Reducing systems of algebraic equations to polynomials is common practice in higher algebra (e.g. see the book [15]). In elimination theory [16] this reduction actually constitutes the solution of the system.

We refer to a polynomial in the steady-state reaction rate as a kinetic polynomial [17–23]. In present work we'll study the properties of the kinetic polynomial for a nonlinear single pathway mechanism of a catalytic reaction with a single type of active sites. Note that as a steady-state kinetic equation for a linear single pathway mechanism and as equations (1.2) and (1.3) for the limiting cases of rate-determining step and neighbourhood of equilibrium this polynomial must satisfy an "equilibrium test", i.e. at $w = 0$ we must have (1.5). On the basis of kinetic polynomial we'll consider classical concepts in the kinetics of complex reactions: "stoichiometric number", "molecularity", "limitation", "equilibrium" and refine their contents and the range of applicability. Then the possibilities of kinetic polynomial application for the solving of direct and inverse kinetic problems as well as the algorithm of computation of this polynomial and it realisation via the computer algebra will be discussed.

2. Theory. The following nonlinear single pathway mechanism of a heterogeneous catalytic reaction is considered:

$$(2.1) \qquad \sum_{r=1}^{l} \gamma'_{ir} C_r + \sum_{j=1}^{n} \alpha_{ij} Z_j \underset{k_{-i}}{\overset{k_i}{\rightleftharpoons}} \sum_{r=1}^{l} \gamma''_{ir} C_r + \sum_{j-1}^{n} \beta_{ij} Z_j,$$

where Z_j and C_r are jth surface and rth observed substances having concentrations z_j and c_r respectively; $\alpha_{ij}, \beta_{ij}, \gamma'_{ir}$, and γ''_{ir} are their stoichiometric coefficients at ith step in the forward and back directions; and k_i and k_{-i} are the rate constants.

The reaction weights are $b_i = k_i \prod_{r=1}^{l} c_r^{\gamma'_{ir}}$ and $b_{-i} = k_{-i} \prod_{r=1}^{l} c_r^{\gamma''_{ir}}$. Stoichiometric coefficients of the surface substances satisfy the following restrictions:

$$(2.2) \qquad \sum_{j=1}^{n} (\alpha_{ij} - \beta_{ij}) = 0, \quad i = 1, \ldots, n$$

i.e. the total number of molecules of surface substances remains unchanged at every step[4]. Besides, we'll consider mechanisms for only one type of active site. According to the steady-state reaction theory (see Eq. (1.1)), the system of quasi-steady-state equations for mechanism (2.1) is of the form

$$(2.3.1) \qquad b_1 \prod_{j=1}^{n} z_j^{\alpha_{1j}} - b_{-1} \prod_{j=1}^{n} z_j^{\beta_{1j}} = \nu_1 w,$$

$$(2.3.2) \qquad b_2 \prod_{j=1}^{n} z_j^{\alpha_{2j}} - b_{-2} \prod_{j=1}^{n} z_j^{\beta_{2j}} = \nu_2 w,$$

$$\vdots \qquad \vdots$$

$$(2.3.\text{n}) \qquad b_n \prod_{j=1}^{n} z_j^{\alpha_{nj}} - b_{-n} \prod_{j=1}^{n} z_j^{\beta_{nj}} = \nu_n w,$$

$$(2.3.\text{n+1}) \qquad \sum_{j=1}^{n} z_j = 1;$$

where ν_i is the stoichiometric number of the ith step (here the set ν_i can be chosen arbitrary, provided that adding up the steps multiplied by respective ν_i results in a net equation free from intermediates), w is the steady-state reaction rate.

Let $\nu_1 \neq 0$. Let us split system (2.3) into Eq. (2.3.1) and the subsystem (2.3.2)– (2.3.n+1). Suppose that at a fixed value of w the subsystem has a finite number of solutions $\langle z_{1(r)}(w), \ldots, z_{n(r)}(w) \rangle$ with $r = 1, \ldots, M^5$; the solutions are taken with their multiplicities. For a further analysis we need the following function

$$(2.4) \qquad R(w) = \prod_{r=1}^{M} \left(b_1 \prod_{j=1}^{n} z_{j(r)}^{\alpha_{1j}}(w) - b_{-1} \prod_{j=1}^{n} z_{j(r)}^{\beta_{1j}}(w) - \nu_1 w \right).$$

$R(w)$ is the resultant (see [24]) of Eq. (2.3.1) with respect to subsystem (2.3.2)– (2.3.n+1). (Definition of resultant in the classic theory of elimination is given in [15,16]). The resultant of a system of algebraic equations is a rational function (i.e. polynomial or a ratio of two polynomials) [24], but in our case function $R(w)$ defined by formulae (2.4) is namely polynomial (for proof see works [22,23]):

$$(2.5) \qquad R(w) = B_L w^L + \cdots + B_1 w + B_0.$$

---

[4] Restrictions (2.2) correspond to mechanisms (2.1) where each surface substance contains the same number of active sites. The case most commonly encountered is a surface substance containing only one active site.

[5] It was proved that if $w = 0$ is not a root of system (2.3) and $\nu_1 \neq 0$ then subsystem like (2.3.2)–(2.3.n+1) has only discrete roots in C^n and has no roots at infinite hyperplane [22,23]. This assumption means that system (2.3) has no boundary steady states (see below).

The resultant $R(w)$ is zero for these and only these values of w which are solutions to system (2.3).

The function $R(w)$ is a kinetic polynomial. Its coefficients $B_0, \ldots, B_L$ can be written in the form of polynomials in b_i and b_{-i}. We will write $B_i \sim a$ if $B_i/a \neq 0$ when $b_{\pm 1}, \ldots, b_{\pm n} \neq 0$. Consider now the properties of B_i.

THEOREM 2.1.

(2.6) $$B_0 \sim (b_1^{\nu_1} \ldots b_n^{\nu_n} - b_{-1}^{\nu_1} \ldots b_{-n}^{\nu_n})^p$$

where stoichiometric numbers $\nu_1, \ldots, \nu_n$ have no greatest common divisor greater than 1 and p is positive integer.

Proof. It follows from (2.4) and (2.5) that

(2.7) $$B_0 = \prod_{r=1}^{M} \left(b_1 \prod_{j=1}^{n} z_{j(r)}^{\alpha_{1j}}(0) - b_{-1} \prod_{j=1}^{n} z_{j(r)}^{\beta_{1j}}(0) \right).$$

The solutions $\mathbf{z}_r(0)$ of subsystem (2.3.2)–(2.3.n+1) for $w = 0$ may be interior and boundary value. An interior root is such that all $z_{j(r)}$ are non zero. Boundary solutions exist if $z_{j(r)}$ are zero for certain j. It is easy to show (see [23]) that when $\nu_{k+1}, \ldots, \nu_n = 0$ and $\nu_1 \neq 0, \ldots, \nu_k \neq 0$ we must take into account only such roots $\mathbf{z}_r(0)$ that have k or more non-zero coordinates.

Consider at first interior roots. After substitution $z_j(0) = \rho_j e^{\mathfrak{F}\phi_j}$, with $\mathfrak{F}^2 = -1$ we'll get

(2.8.1)
$$\prod_{j=1}^{n-1} \rho_j^{\alpha_{2j} - \beta_{2j}} = (b_{-2}/b_2)\rho_n^{\beta_{2n} - \alpha_{2n}}$$
$$\vdots \qquad \vdots$$
$$\prod_{j=1}^{n-1} \rho_j^{\alpha_{nj} - \beta_{nj}} = (b_{-n}/b_n)\rho_n^{\beta_{nn} - \alpha_{nn}}$$

(2.8.2)
$$\sum_{j=1}^{n-1} (\alpha_{2j} - \beta_{2j})\phi_j = (\beta_{2n} - \alpha_{2n})\phi_n + 2\pi m_2$$
$$\vdots \qquad \vdots$$
$$\sum_{j=1}^{n-1} (\alpha_{nj} - \beta_{nj})\phi_j = (\beta_{nn} - \alpha_{nn})\phi_n + 2\pi m_n$$

(2.8.3)
$$\sum_{j=1}^{n} \rho_j \cos \phi_j = 1$$
$$\sum_{j=1}^{n} \rho_j \sin \phi_j = 0$$

with $m_j = 0, \pm 1, \pm 2, \ldots$ (the m_j must be chosen so as to obtain all possible solutions $\langle z_i(0) \rangle$). Taking logs in (2.8.1) we get

$$(2.9) \qquad \log \rho_i = \left(\sum_{j=2}^{n} A_{ji}(\log(b_{-j}/b_j) + (\beta_{jn} - \alpha_{jn})\log \rho_n) \right) / \Delta_1$$

where

$$\Delta_1 = \det \begin{pmatrix} (\alpha_{21} - \beta_{21}) & \cdots & (\alpha_{2,n-1} - \beta_{2,n-1}) \\ \vdots & \ddots & \vdots \\ (\alpha_{n1} - \beta_{n1}) & \cdots & (\alpha_{n,n-1} - \beta_{n,n-1}) \end{pmatrix}$$

and A_{ji} are the cofactors of the elements in Δ_1 with the corresponding indices. If $\Delta_1 = 0$, which is possible, for instance, in the presence of "buffer" steps, another subsystem can be chosen. All Δ_i cannot be zero, for in our instance the stoichiometric matrix is of rank $n - 1$). Expression (2.9) can be transformed thus:

$$(2.10) \qquad \log \rho_i = \left(\sum_{j=2}^{n} A_{ji} \log(b_{-j}/b_j) + \log \rho_n \sum_{j=2}^{n} A_{ji}(\beta_{jn} - \alpha_{jn}) \right) / \Delta_1$$

From (2.2) we have

$$\sum_{j=2}^{n} A_{ji}(\beta_{jn} - \alpha_{jn}) = \det \begin{pmatrix} (\alpha_{21} - \beta_{21}) & \cdots & \sum_{l=1}^{n-1}(\alpha_{2l} - \beta_{2l}) & \cdots & (\alpha_{2,n-1} - \beta_{2,n-1}) \\ \vdots & & \vdots & & \vdots \\ (\alpha_{n1} - \beta_{n1}) & \cdots & \sum_{l=1}^{n-1}(\alpha_{nl} - \beta_{nl}) & \cdots & (\alpha_{n,n-1} - \beta_{n,n-1}) \end{pmatrix}$$

This determinant is obtained from Δ_1 by replacing the ith column with the sum of all columns in Δ_1 and is equal to Δ_1. Then we have

$$(2.11) \qquad \rho_i = \rho_n \prod_{j=2}^{n} (b_{-j}/b_j)^{A_{ji}/\Delta_1}.$$

Similarly, from (2.8.2)

$$(2.12) \qquad \phi_i = \phi_n + (2\pi/\Delta_1) \sum_{j=2}^{n} m_j A_{ji}.$$

From (2.11) and (2.12) for $i = 1, \ldots, n - 1$ we find that

$$(2.13) \qquad z_i(0) = z_n(0) \left(\prod_{j=2}^{n} (b_{-j}/b_j)^{A_{ji}/\Delta_1} \right) \exp \left(2\pi \mathfrak{I} \sum_{j=2}^{n} m_j A_{ji}/\Delta_1 \right)$$

Let us return to relation (2.7). It is obvious that

$$(2.14) \qquad B_0 \sim \Psi = b_{-1}^{M_{in}} \prod_{r=1}^{M_{in}} \prod_{j=1}^{n} z_{j(r)}^{\beta_{1j}}(0)\Psi',$$

where

(2.15)
$$\Psi' = \prod_{r=1}^{M_{in}} \left((b_1/b_{-1}) \prod_{j=1}^{n} z_{j(r)}^{(\alpha_{ij}-\beta_{ij})}(0) - 1 \right).$$

Here M_{in} is the number of interior solutions of the subsystem. It was proved [23] on the basis of results [25] that in our case

(2.16)
$$M_{in} = |\Delta_1|$$

and these roots are simple. Substituting (2.13) into (2.15), we obtain

(2.17)
$$\Psi' = \prod_{r=1}^{M_{in}} \left(\frac{b_1}{b_{-1}} \left(\frac{b_{-2}}{b_2} \right)^{\Delta_2/\Delta_1} \cdots \left(\frac{b_{-n}}{b_n} \right)^{\Delta_n/\Delta_1} \exp(2\pi\mathfrak{F}(m_2\Delta_2 + \ldots m_n\Delta_n)/\Delta_1) - 1 \right),$$

where $\Delta_i, i = 2, \ldots, n$, are the determinants obtained from Δ_1 by replacing row i with the row $\langle (\alpha_{11} - \beta_{11}) \ldots (\alpha_{1,n-1} - \beta_{1,n-1}) \rangle$. We consider the sum

(2.18)
$$S = \Delta_1(\alpha_{1j} - \beta_{1j}) - \Delta_2(\alpha_{2j} - \beta_{2j}) - \cdots - \Delta_n(\alpha_{nj} - \beta_{nj}).$$

On the other hand

(2.19)
$$S = \det \begin{pmatrix} (\alpha_{1j} - \beta_{1j}) & (\alpha_{11} - \beta_{11}) & \cdots & (\alpha_{1,n-1} - \beta_{1,n-1}) \\ \vdots & \vdots & \ddots & \vdots \\ (\alpha_{nj} - \beta_{nj}) & (\alpha_{n1} - \beta_{n1}) & \cdots & (\alpha_{n,n-1} - \beta_{n,n-1}) \end{pmatrix}.$$

Then the cofactors of the elements in the first row are $\Delta_1, -\Delta_2, \ldots, -\Delta_n$. However, determinant (2.19) is zero as it contains the same columns for $j \neq n$ and is the nth-order determinant of a matrix of rank $n-1$ for $j = n$. Consequently, S is identically zero, and $\Delta_1, -\Delta_2, \ldots, -\Delta_n$ can be considered as the stoichiometric numbers of the corresponding steps in mechanism (2.1). Equation (2.17) contains Δ_i divided by Δ_1. Reducing $(m_2\Delta_2 + \ldots m_2\Delta_n)/\Delta_1$ in (2.17) by the greatest common divisor, we obtain

(2.20)
$$-(m_2\Delta_2 + \ldots m_n\Delta_n)/\Delta_1 \equiv (m_2\nu_2 + \ldots m_n\nu_n)/\nu_1$$

The number set $\{m_2, \ldots, m_n\}$ must contain a certain minimum set that yields different values of $\exp(-2\pi\mathfrak{F}\sum_{i=2}^{n} m_i\nu_i/\nu_1)$ i.e. all different values of remainders from division $- \sum_{i=2}^{n} m_i\nu_i/\nu_1$. Let ν_1 exceed zero. There exist such set $m_1, \ldots, m_n$, that $-(m_1\nu_1 + \cdots + m_n\nu_n) = 1$ because $\nu_1, \ldots, \nu_n$ have no common divisor other then unity. Then, for obtaining all possible remainders it's sufficient to multiply $m_1, \ldots, m_n$ by $\nu_1 - l$, where $l = 1, \ldots, \nu_1$.

It follows from (2.17), (2.20) for this way obtained sets $\{\mathbf{m}\}$ that

(2.21)
$$\Psi' \sim \prod_{s=0}^{\nu_1-1} ((b_1/b_{-1})(b_2/b_{-2})^{\nu_2/\nu_1} \ldots (b_n/b_{-n})^{\nu_n/\nu_1} \exp(2\pi\mathfrak{F}s/\nu_1) - 1)$$
$$\equiv (-1)^{\nu_1-1} \left(\prod_{i=1}^{n} (b_i/b_{-i})^{\nu_i} - 1 \right)$$

Thus, we have proved that

$$
(2.22) \qquad \prod_{r=1}^{M_1 n} \left(b_1 \prod_{j=1}^{n} z_{j(r)}^{\alpha_{1j}}(0) - b_{-1} \prod_{j=1}^{n} z_{j(r)}^{\beta_{1j}}(0) \right) \sim \left(\prod_{i=1}^{n} (b_i/b_{-i})^{\nu_i} - 1 \right)^p
$$

(about degree p see below).

Let now $\nu_n = 0$. Consider the roots (if they are exist) that have one zero coordinate, for instance, $z_1 \neq 0, \ldots, z_{n-1} \neq 0$ and $z_n = 0$. If the product $\prod_{j=1}^{n} z_j^{\alpha_{1j}}$ or the product $\prod_{j=1}^{n} z_j^{\beta_{1j}}$ includes z_n then these roots doesn't influence the result (2.22). If both these products doesn't include z_n then assuming $z_n = 0$ we get the $(n-1)$-dimensional system like (2.8) and relation (2.22) is fulfilled in this case too. $\square$

COROLLARY 2.1. *If $\nu_1 \neq 0, \ldots, \nu_n \neq 0$ then the degree p in (2.6) is the greatest common divisor of $\Delta_i : p = |\Delta_1/\nu_1| = \cdots = |\Delta_n/\nu_n|$.*

Proof. Number of interior roots is $|\Delta_1|$ and $|\nu_1|$ of these roots gives one multiplier $(\prod_{i=1}^{n} (b_i/b_{-i})^{\nu_i} - 1)^6$. $\square$

COROLLARY 2.2. *If degree $p > 1$ and $\nu_1, \ldots, \nu_n \neq 0$ then coefficient B_s of $R(w)$ includes the multiplier (the cyclic characteristic) $C = (b_1^{\nu_1} \ldots b_n^{\nu_n} - b_{-1}^{\nu_1} \ldots b_{-n}^{\nu_n})$ with degree $p - s, s = 0, 1, \ldots, p-1; p \leq L$, where L is degree of kinetic polynomial.*

Proof. It follows from (2.4) and (2.5) that first coefficient of $R(w)$ is

$$
(2.23) \qquad B_1 = R'(w)|_{w=0} = \sum_{j} (u_1'(z_{(j)}(0)) - \nu_1) \prod_{i \neq j} u_1(z_{(i)}(0)),
$$

where $u_1(z_{(j)}(0))$ is the value of the rate of 1st step in the solution $\langle z_{1(j)}(0), \ldots, z_{n(j)}(0) \rangle$.

We can see from Theorem 2.1 and Corollary 2.1 that $|\Delta_1|$ interior roots of subsystem can be divided to p groups contained ν_1 elements and each one produces multiplier C in the product (2.7).

Thus $\prod_{i \neq j} u_1(z_{(i)}(0)) \sim C^{p-1}$. For the second derivative the degree of C is two units smaller and so on. The degree of kinetic polynomial $L \geq \max_i |\Delta_i| \geq |\Delta_i/\nu_i| = p$, so multiplier C doesn't contained in B_L. $\square$

Consider now the form of the other coefficients of kinetic polynomial. It follows from (2.5) that

$$
(2.24) \qquad \frac{B_i}{B_0} = \frac{1}{i! R(0)} \left. \frac{d^i R(w)}{dw^i} \right|_{w=0}, \qquad i = 1, \ldots, L,
$$

where $R(w)$ is defined by (2.4). The recursive form of (2.24) is

$$
(2.25) \qquad \frac{B_i}{B_0} = \frac{1}{i} \sum_{k=0}^{i-1} \frac{1}{k!} \frac{B_{i-1-k}}{B_0} \left. \frac{d^{k+1} \ln R(w)}{dw^{k+1}} \right|_{w=0}
$$

---

[6] If, for instance, $\nu_n = 0$ and there are exist the roots with $z_n = 0$ and $\prod_{j=1}^{n} z_j^{\alpha_{1j}} \neq 0$ and $\prod_{j=1}^{n} z_j^{\beta_{1j}} \neq 0$ then the number of these roots is $|\Delta_1|/|\alpha_{nn} - \beta_{nn}|$. Thus the degree p increased by $|\Delta_1/(\nu_1(\alpha_{nn} - \beta_{nn}))|$ (for details see [23]).

It follows, in particular, from (2.25) that

$$(2.26) \qquad \frac{B_1}{B_0} = \frac{d\ln R(w)}{dw}\bigg|_{w=0},$$

and

$$(2.27) \qquad \frac{B_2}{B_0} = \frac{1}{2}\left(\left(\frac{B_1}{B_0}\right)^2 + \frac{d^2\ln R(w)}{dw^2}\bigg|_{w=0}\right).$$

Thus to find the coefficients of (2.5) we must calculate derivatives of $\ln R(w)$. Let's prove that

$$(2.28) \qquad \frac{d\ln R(w)}{dw} = -\sum_{k=1}^{n}\nu_k\sum_{j_k=1}^{M_k}\frac{1}{u_k(\mathbf{z}_{jk}(w)) - \nu_k w}$$

where $u_k(\mathbf{z}_{jk}(w))$ is the rate of kth step calculated in the solution of subsystem obtained from (2.3) by elimination of kth equation (2.3.1–2.3.n), M_k is the number of roots of kth subsystem[7]. Using the formulae for implicit function derivative we get from (2.4)
(2.29)

$$\frac{d\ln R(w)}{dw} = -\nu_1\sum_{j_1=1}^{M_1}\frac{1}{u_1(z_{j1}(w)) - \nu_1 w} + \sum_{k=2}^{n}\nu_k\sum_{j_1=1}^{M_1}\frac{(-1)^k J_k(\mathbf{z}_{j1}(w))}{J_1(z_{j1}(w))(u_1(z_{j1}(w)) - \nu_1 w)},$$

where $J_1(\cdot)$, $J_k(\cdot)$ are the Jacobians of corresponding subsystems. Let now prove the identity

$$(2.30) \qquad \sum_{j_1=1}^{M_1}\frac{(-1)^k J_k(z_{j1}(w))}{J_1(z_{j1}(w))(u_1(z_{j1}(w)) - \nu_1 w)} \equiv -\sum_{j_k=1}^{M_k}\frac{1}{u_k(z_{jk}(w)) - \nu_k w}$$

Consider the following system, depending on the parameter w

$$(2.31.1) \qquad (u_1(z) - \nu_1 w)(u_k(z) - \nu_k w) = 0,$$
$$(2.31.2) \qquad u_2(z) - \nu_2 w = 0,$$

$$\vdots \qquad \vdots$$

$$u_{k-1}(z) - \nu_{k-1} w = 0,$$
$$u_{k+1}(z) - \nu_{k+1} w = 0,$$
$$(2.31.n) \qquad u_n(z) - \nu_n w = 0,$$
$$(2.31.n+1) \qquad \sum_{i=1}^{n} z_i = 1$$

This system has following properties:

(1) The C^n into C^n mapping given by lhs of (2.31) is polynomial mapping,

---

[7]It's sufficient to consider in (2.28) the values of w in the neighbourhood of zero.

(2) It follows from (2.2) that $u_j(z)$ are homogeneous polynomials and one can prove [22,23] that in our assumptions the system $u_i(z) = 0$, $(i \neq k)$, $\sum_{i=1}^{n} z_i = 0$ has unique solution $z_1 = \cdots = z_n = 0$. Thus, the higher degree homogeneous components of equations (2.31) have a single common zero point 0,

(3) The degree of system (2.31) Jacobian is $\deg J = \sum_{i=1}^{n} \deg u_i + 1 - n$. The degree of Jacobian J_k of the system $u_i(z) = 0$, $(i \neq k)$, $\sum_{i=1}^{n} z_i = 1$ is $\deg J_k = \sum_{i=1}^{n} \deg u_i + 1 - n - \deg u_k$. Thus $\deg J_k < \deg J$.

It follows from the facts (1)–(3) the applicability of Euler-Jacobi formulae [26], i.e.

$$(2.32) \qquad \sum_{j_{1k}=1}^{M_{1k}} \frac{J_k(z_{j_{1k}}(w))}{J(z_{j_{1k}}(w))} \equiv 0,$$

where $M_{1k} = M_1 + M_k$, and

$$(2.33) \qquad \begin{aligned} J &= (u_1(z(w)) - \nu_1 w)(-1)^k J_1(z(w)) + (u_k(z(w)) - \nu_k w)J_k(z(w)) \\ &= \begin{cases} (u_1(z(w)) - \nu_1 w)(-1)^k J_1(z(w)), & z(w) = z_{j1}(w) \\ (u_k(z(w)) - \nu_k w)J_k(z(w)), & z(w) = z_{jk}(w) \end{cases} \end{aligned}$$

Then from (2.32), (2.33) we have

$$(2.34)$$
$$\sum_{j_1=1}^{M_1} \frac{J_k(z_{j1}(w))}{(-1)^k J_1(z_{j1}(w))(u_1(z_{j1}(w)) - \nu_1 w)} + \sum_{j_k=1}^{M_k} \frac{J_k(z_{jk}(w))}{(u_k(z_{jk}(w)) - \nu_k w)J_k(z_{j1}(w))} \equiv 0$$

The identities (2.34) and (2.30) are identical. Then from (2.29) and (2.30) we get the relation (2.28).

THEOREM 2.2.

$$(2.35) \qquad \frac{B_1}{B_0} = -\sum_{k=1}^{n} \nu_k \sum_{j_k=1}^{M_k} \frac{1}{u_k(z_{j_k}(0))},$$

where M_k is the number of solutions (with their multiplicities) of the kth equilibrium subsystem obtained by setting $w = 0$ in Eqs. (2.3.1)–(2.3.n), except the kth equation, and $u_k(z_{j_k}(0))$ are the values of the rate of the kth step in the solution $z_{j_k}(0) = \langle z_{1_{j_k}}(0), \ldots, z_{n_{j_k}}(0) \rangle$ of kth equilibrium subsystem.

Proof. Formulae (2.35) follows directly from (2.26) and (2.28). □

Differentiation of (2.28) enables to find the relations for higher coefficients[8].

THEOREM 2.3.

$$(2.36) \qquad \frac{B_2}{B_0} = \frac{1}{2}\left(\left(\frac{B_1}{B_0}\right)^2 - \sum_{s,k=1}^{n} \sum_{j_k=1}^{M_k} (-1)^{s+k} \nu_s \nu_k \frac{J_s}{J_k u_k^2}\bigg|_{z=z_{j_k}(0)} \right)$$

---

[8] Another theory based on the results of complex analysis (multidimensional logarithmic residue theory) is given in [22,23].

Proof. From (2.28) we get

$$(2.37) \qquad \frac{d^2 \ln R(w)}{dw^2} = \sum_{k=1}^{n} \nu_k \sum_{j_k=1}^{M_k} \frac{u'_k - \nu_k}{(u_k - \nu_k w)^2},$$

where

$$(2.38) \qquad u'_k = (-1/J_k) \sum_{s \neq k}^{n} (-1)^{s+k} \nu_s J_s \big|_{z=z_{j_k}(w)}$$

Then from (2.37)

$$(2.39) \qquad \frac{d^2 \ln R(w)}{dw^2} = -\sum_{k=1}^{n} \sum_{s=1}^{n} \nu_k \nu_s \sum_{j_k=1}^{M_k} \frac{(-1)^{s+k} J_s(z_{jk}(w))}{J_k(z_{jk}(w))(u_k(z_{jk}(w)) - \nu_k w)^2}$$

Thus, from relations (2.27) and (2.39) we get (2.36). □

Remark 2.1. The following identity is valid ([27])

$$(2.40) \qquad \sum_{j_l=1}^{M_l} \frac{J_s}{J_l u_l^2} \bigg|_{z=z_{jl}(0)} \equiv \sum_{j_s=1}^{M_s} \frac{J_l}{J_s u_s^2} \bigg|_{z=z_{js}(0)}$$

Remark 2.2. $u_k(z_{jk}(0)) \neq 0$ from the assumption that system (2.3) has no boundary steady states.

Remark 2.3. We can use (2.36) when equilibrium subsystems have multiple roots. This fact follows from the corollaries of theorem about inverse Jacobian [26].

Thus the coefficients of kinetic polynomial, calculated by formulae (2.25), (2.28), are the symmetric functions of the equilibrium subsystem solutions $z_{jk}(0)$. One can use (2.13) when finding $z_{jk}(0)$.

Here are some results and physicochemical corollaries.

(1) A system of nonlinear quasi-steady-state equations for a single pathway mechanism can always be reduced to a polynomial in terms of the steady-state reaction rate. The polynomial coefficients are polynomials in the kinetic parameters. The lowest coefficient is proportional to the cyclic characteristic $C = (b_1^{\nu_1} \ldots b_n^{\nu_n} - b_{-1}^{\nu_1} \ldots b_{-n}^{\nu_n})$. After substitution of expressions for $b_{\pm i}$ into C we obtain

$$(2.41) \qquad C = \prod_{i=1}^{n} k_i^{\nu_i} \prod_{r=1}^{l} c_r^{\Gamma_{rI}} - \prod_{i=1}^{n} k_{-i}^{\nu_i} \prod_{r=1}^{l} c_r^{\Gamma_{rT}},$$

where $\Gamma_{rI} = \sum_{i=1}^{n} \nu_i \gamma'_{ir}$ and $\Gamma_{rT} = \sum_{i=1}^{n} \nu_i \gamma''_{ir}$ are the stoichiometric coefficients of observed species in the net chemical equation.

It follows from (2.5) that at equilibrium $(w = 0)B_0 = 0$ and

$$K = \prod_{i=1}^{n} k_i^{\nu_i} / \prod_{i=1}^{n} k_{-i}^{\nu_i} = \prod_{r=1}^{l} c_r^{\Gamma_{r}T} / \prod_{r=1}^{l} c_r^{\Gamma_{r}I}$$

(see (1.5)). Thus such a form of the lowest coefficient ensures the thermodynamic validity of kinetic polynomial.

(2) Relation (2.6) contains a set of values ν_i that have no common divisor other than unity. This set is determined from reaction mechanism (2.1). To find the set ν_i it's sufficient to calculate the determinants $\Delta_1, -\Delta_2, \ldots, -\Delta_n$ and eliminate their greatest common divisor p. The meaning of these determinants is simple: they are the cofactors N_i of an arbitrary column of the stoichiometric matrix for the intermediates in mechanism (2.1).

Thus the stoichiometric numbers contained in B_0 form a minimum set of integers. The arbitrary nature of sets of stoichiometric numbers has been underscored more than once in the literature (e.g. see [28]). However, the set in B_0 is not arbitrary: it depends on the detailed mechanism and defines the form of kinetic low of complex reaction (2.1). The net equation corresponding to this set of stoichiometric numbers does not necessarily have a minimum values of stoichiometric coefficients of the reactants: the values of Γ_{rI}, Γ_{rT} in (2.41) can have greatest common divisor other than unity (it was called "multiplicity" in [14]). For instance, for the simple linear mechanism

$$1)A + Z \rightleftharpoons AZ, \quad 2)A + AZ \rightleftharpoons A_2Z, \quad 3)A_2Z \rightleftharpoons 2B + Z$$

the set $\nu_i\langle 1, 1, 1\rangle$ yields the net equation $2A \rightleftharpoons 2B$, which, in contrast to the equation $A \rightleftharpoons B$ is physically meaningful, namely, it corresponds to the reaction cyclic characteristic $k_1 k_2 k_3 c_A^2 - k_{-1} k_{-2} k_{-3} c_B^2$.

(3) It follows from corollaries 2.1 and 2.2 that when $p = |N_i/\nu_i| > 1$ kinetic polynomial takes the form

(2.42) $$B_L w^L + \cdots + B_p w^p + B'_{p-1} C w^{p-1} + \cdots + B'_0 C^p = 0$$

The value of p defines the dimension of equilibrium neighbourhood (see below) i.e. we must take into account only coefficients $B_p, \ldots, B'_0$ in this case.

(4) It follows from (2.35) that for linear mechanism the inverse of the steady-state reaction rate is equal to the sum of the inverted reaction rates calculated assuming that the $1, 2, \ldots, n$th steps are rate-limiting and the remaining steps are at equilibrium:

(2.43) $$\frac{1}{w} = \sum_{i=1}^{n} \frac{1}{w_i}$$

It is easy to get from (2.43) the equation (1.4).

(5) The solution of the kinetic polynomial cannot generally be represented in the explicit form like (1.2), (1.3) customary to the chemist. These equations are valid only in the following cases: in the neighbourhood equilibrium (a), when there exist the rate-limiting step (b), and, for a linear mechanism (c).

3. Examples. Let's consider the impact mechanism

(3.1)

$$1) \quad A_2 + 2Z \underset{k_{-1}}{\overset{k_1}{\rightleftharpoons}} 2AZ \qquad |1|$$

$$2) \quad AZ + B \underset{k_{-2}}{\overset{k_2}{\rightleftharpoons}} AB + 2Z \quad |2|$$

$$\overline{A_2 + 2B \rightleftharpoons 2AB}$$

where A_2, B, and AB are observed species, and Z, AZ are the surface intermediates. After elementary transformations we'll get kinetic polynomial

$$(3.2) \quad 4(b_1 - b_{-1})w^2 - (4(b_1 b_2 + b_{-1} b_{-2}) + (b_2 + b_{-2})^2)w + (b_1 b_2^2 - b_{-1} b_{-2}^2) = 0,$$

where $b_1 = k_1 c_{A_2}, b_{-1} = k_{-1}, b_2 = k_2 c_B, b_{-2} = k_{-2} c_{AB}$. The form of the free term corresponds to (2.6) at $p = 1$. Its exponents correspond to the minimal-integer stoichiometric numbers $\langle 1, 2 \rangle$. Physically meaningful solution of (3.2) can be written as:

(3.3)

$$w = \frac{2(b_1 b_2^2 - b_{-1} b_{-2}^2)}{4(b_1 b_2 + b_{-1} b_{-2}) + (b_2 + b_{-2})^2 + (b_2 + b_{-2})\sqrt{(b_2 + b_{-2})^2 + 8(b_1 b_2 + b_{-1} b_{-2}) + 16 b_{-1} b_1}}$$

The form of (3.3) corresponds to eq. (1.4): the numerator contains cyclic characteristic and denominator is the sum of positive reaction weight functions.

For mechanism

(3.4)

$$1) \quad A_2 + 2Z \underset{k_{-1}}{\overset{k_1}{\rightleftharpoons}} 2AZ \qquad |1|$$

$$2) \quad 2AZ \underset{k_{-2}}{\overset{k_2}{\rightleftharpoons}} B_2 + 2Z \quad |1|$$

$$\overline{A_2 \rightleftharpoons B_2}$$

we have $N_1 = N_2 = 2, \nu_1 = \nu_2 = 1, p = |N_i/\nu_i| = 2$. In accordance with corollaries 2.1 and 2.2 (see, also eq. (2.42)) we get

(3.5)

$$(b_{-1} + b_2 - b_{-2} - b_1)^2 w^2 - 2(b_1 b_2 - b_{-1} b_{-2})(b_{-1} + b_1 + b_{-2} + b_2)w + (b_1 b_2 - b_{-1} b_{-2})^2 = 0,$$

where $b_1 = k_1 c_{A_2}, b_{-1} = k_{-1}, b_2 = k_2, b_{-2} = k_{-2} c_{B_2}$.

Consider now the following three-step adsorption mechanism

(3.6)

$$1) \quad A_2 + 2Z \underset{k_{-1}}{\overset{k_1}{\rightleftharpoons}} 2AZ \qquad |1|$$

$$2) \quad B + Z \underset{k_{-2}}{\overset{k_2}{\rightleftharpoons}} BZ \qquad |2|$$

$$3) \quad AZ + BZ \underset{k_{-3}}{\overset{k_3}{\rightleftharpoons}} 2Z + AB \quad |2|$$

$$\overline{A_2 + 2B \rightleftharpoons 2AB}$$

where A_2, B, and AB are observed species, and AZ and BZ are the adsorbed substances, and Z is the active site of the surface. The following system of quasi-steady-state equations corresponds to mechanism (3.6)

$$(3.7.1) \qquad b_1 z_1^2 - b_{-1} z_2^2 = w,$$
$$(3.7.2) \qquad b_2 z_1 - b_{-2} z_3 = 2w,$$
$$(3.7.3) \qquad b_3 z_2 z_3 - b_{-3} z_1^2 = 2w$$
$$(3.7.4) \qquad z_1 + z_2 + z_3 = 1,$$

where $z_1 = [Z], z_2 = [AZ], z_3 = [BZ], b_1 = k_1 c_{A_2}, b_{-1} = k_{-1}, b_2 = k_2 c_B, b_{-2} = k_{-2}, b_3 = k_3, b_{-3} = k_{-3} c_{AB}$. Obtaining $z_3 = (b_2 z_1 - 2w)/b_{-2}$ and $z_2 = 1 - z_1 - z_3$ from eqs. (3.7.2) and (3.7.4) and substituting them into eqs. (3.7.1) and (3.7.3), we reduce eqs. (3.7) to the following system:

$$(3.8.1) \qquad d_{01} z_1^2 + d_{11} z_1 + d_{21} = 0,$$
$$(3.8.2) \qquad d_{02} z_1^2 + d_{12} z_1 + d_{22} = 0,$$

where

$$d_{01} = b_{-1}(b_{-2} + b_2)^2 - b_{-2}^2 b_1,$$
$$d_{11} = -2b_{-1}(b_{-2} + b_2)(b_{-2} + 2w),$$
$$d_{21} = 4b_{-1}w^2 + b_{-2}(b_{-2} + 4b_{-1})w + b_{-1}b_{-2}^2,$$
$$d_{02} = b_2 b_3 (b_{-2} + b_2) + b_{-2}^2 b_{-3},$$
$$d_{12} = -b_3(2(b_{-2} + 2b_2)w + b_{-2}b_2),$$
$$d_{22} = 4b_3 w^2 + 2b_{-2}(b_{-2} + b_3)w$$

To obtain a kinetic polynomial it is sufficient to find the resultant $R(w)$ of polynomials (3.8) for the variable z_1. The result of its computation obtained utilising REDUCE computer algebra system is[9]

$$(3.9) \qquad R(w) = B_4 w^4 + B_3 w^3 + B_2 w^2 + B_1 w + B_0 = 0$$

where

(3.10)
$$B_0 = b_{-1} b_{-2}^2 (b_{-1} b_{-2}^2 b_{-3}^2 - b_1 b_2^2 b_3^2)$$

(3.11)
$$B_1 = 4b_{-1}^2 b_{-2}^4 b_{-3} + 8b_{-1}^2 b_{-2}^3 b_{-3}^2 + 8b_{-2}^2 b_{-2}^3 b_{-3} b_2 + 4b_{-1}^2 b_{-2}^2 b_{-3} b_2^2 + 2b_{-1} b_{-2}^4 b_{-3}^2 +$$
$$4b_{-1} b_{-2}^4 b_{-3} b_1 + 4b_{-1} b_{-2}^3 b_{-3} b_1 b_3 + 2b_{-1} b_{-2}^3 b_{-3} b_2 b_3 + 8b_{-1} b_{-2}^3 b_1 b_2 b_3 + 2b_{-1} b_{-2}^2 b_{-3} b_2^2 b_3 +$$
$$8b_{-1} b_{-2}^2 b_1 b_2^2 b_3 + 4b_{-1} b_{-2}^2 b_1 b_2 b_3^2 + b_{-1} b_{-2}^2 b_2^2 b_3^2 - 4b_{-1} b_{-2} b_1 b_2^2 b_3^2 + 2b_{-1} b_{-2} b_2^3 b_3^2 +$$
$$b_{-1} b_2^4 b_3^2 - b_{-2}^2 b_1 b_2^2 b_3^2$$

---

[9] The factor b_{-2}^4 contained in $B_0, \ldots, B_4$ was cancelled.

(3.12)
$$B_2 = 4b_{-1}^2 b_{-2}^4 + 16b_{-1}^2 b_{-2}^3 b_{-3} + 16b_{-1}^2 b_{-2}^3 b_2 + 24b_{-1}^2 b_{-2}^2 b_{-3}^2 + 32b_{-1}^2 b_{-2}^2 b_{-3} b_2 +$$
$$24b_{-1}^2 b_{-2}^2 b_2^2 + 16b_{-1}^2 b_{-2} b_{-3} b_2^2 + 16b_{-1}^2 b_{-2} b_2^3 + 4b_{-1}^2 b_2^4 - 4b_{-1} b_{-2}^4 b_{-3} -$$
$$8b_{-1} b_{-2}^4 b_1 + 8b_{-1} b_{-2}^3 b_{-3}^2 + 16b_{-1} b_{-2}^3 b_{-3} b_1 - 8b_{-1} b_{-2}^3 b_{-3} b_2 - 8b_{-1} b_{-2}^3 b_{-3} b_3 -$$
$$16b_{-1} b_{-2}^3 b_1 b_2 - 8b_{-1} b_{-2}^3 b_1 b_3 - 4b_{-1} b_{-2}^3 b_2 b_3 + 24b_{-1} b_{-2}^2 b_{-3} b_1 b_3 - 4b_{-1} b_{-2}^2 b_{-3} b_2^2 -$$
$$8b_{-1} b_{-2}^2 b_{-3} b_2 b_3 - 8b_{-1} b_{-2}^2 b_1 b_2^2 + 16b_{-1} b_{-2}^2 b_1 b_2 b_3 - 4b_{-1} b_{-2}^2 b_1 b_3^2 - 12b_{-1} b_{-2}^2 b_2^2 b_3 -$$
$$4b_{-1} b_{-2}^2 b_2 b_3^2 + 24b_{-1} b_{-2} b_1 b_2^2 b_3 + 16b_{-1} b_{-2} b_1 b_2 b_3^2 - 12b_{-1} b_{-2} b_2^3 b_3 - 8b_{-1} b_{-2} b_2^2 b_3^2 -$$
$$4b_{-1} b_1 b_2^2 b_3^2 - 4b_{-2}^4 b_3 - 4b_{-1} b_2^3 b_3^2 + b_{-2}^4 b_{-3}^2 + 4b_{-2}^4 b_{-3} b_1 +$$
$$4b_{-2}^4 b_1^2 + 4b_{-2}^3 b_{-3} b_1 b_3 + 2b_{-2}^3 b_{-3} b_2 b_3 + 8b_{-2}^3 b_1^2 b_3 + 4b_{-2}^3 b_1 b_2 b_3 +$$
$$2b_{-2}^2 b_{-3} b_2^2 b_3 + 4b_{-2}^2 b_1^2 b_3^2 + 4b_{-2}^2 b_1 b_2^2 b_3 + b_{-2}^2 b_2^2 b_3^2 - 4b_{-2} b_1 b_2^2 b_3^2 +$$
$$2b_{-2} b_2^3 b_3^2 + b_1^4 b_3^2$$

(3.13)
$$B_3 = 4(4b_{-1}^2 b_{-2}^2 b_{-3} + 8b_{-1}^2 b_{-2} b_{-3}^2 + 8b_{-1}^2 b_{-2} b_{-3} b_2 + 4b_{-1}^2 b_{-3} b_2^2 + 2b_{-1} b_{-2}^2 b_{-3}^2 +$$
$$4b_{-1} b_{-2}^2 b_{-3} b_1 - 4b_{-1} b_{-2}^2 b_{-3} b_3 - 4b_{-1} b_{-2}^2 b_1 b_3 + b_{-1} b_{-2}^2 b_3^2 + 12b_{-1} b_{-2} b_{-3} b_1 b_3 -$$
$$6b_{-1} b_{-2} b_{-3} b_2 b_3 - 4b_{-1} b_{-2} b_1 b_3^2 + 2b_{-1} b_{-2} b_2^2 b_3 - 2b_{-1} b_{-3} b_2^2 b_3 + 4b_{-1} b_1 b_2^2 b_3 +$$
$$4b_{-1} b_1 b_2 b_3^2 + b_{-1} b_2^2 b_3^2 + 2b_{-2}^2 b_{-3} b_1 b_3 + 4b_{-2}^2 b_1^2 b_3 - b_{-2}^2 b_1 b_3^2 +$$
$$4b_{-2} b_1^2 b_3^2 - 2b_{-2} b_1 b_2 b_3^2 - 2b_1 b_2^2 b_3^2)$$

(3.14)
$$B_4 = 16(b_{-1}^2 b_{-3}^2 + 2b_{-1} b_{-3} b_1 b_3 - b_{-1} b_1 b_3^2 + b_1^2 b_3^2)$$

Note that in contrast to the linear mechanism, the cyclic characteristic may "vanish" when some reaction step is irreversible (the first or second step in mechanism (3.6)).

To compute the coefficients of $R(w)$ by formulae (2.35), (2.36) we must find the roots of equilibrium subsystems. 1st subsystem has two solutions ($M_1 = 2$):

(3.15)
$$z_{1(1)} = 0, \quad z_{2(1)} = 1, \quad z_{3(1)} = 0;$$
$$z_{1(2)}/b_{-2} b_2 b_3 = z_{2(2)}/b_{-2}^2 b_{-3} = z_{3(2)}/b_2^2 b_3 = 1/(b_{-2} b_2 b_3 + b_{-2}^2 b_{-3} + b_2^2 b_3)$$

For 2nd subsystem $M_2 = 4$ and

(3.16)
$$z_{1(1,2)} = 0, \quad z_{2(1,2)} = 0, \quad z_{3,(1,2)} = 1;$$
$$z_{1(3,4)}/\pm b_3 \sqrt{b_1 b_{-1}} = z_{2(3,4)}/b_1 b_3 = z_{3(3,4)}/b_{-1} b_{-3} = 1/(\pm b_3 \sqrt{b_1 b_{-1}} + b_1 b_3 + b_{-1} b_{-3})$$

For 3rd subsystem $M_3 = 2$ and

(3.17)
$$z_{1(1,2)}/b_{-2} \sqrt{b_{-1}} = z_{2(1,2)}/\pm b_{-2} \sqrt{b_1} = z_{3(1,2)}/b_2 \sqrt{b_{-1}} = 1/(b_{-2} \sqrt{b_{-1}} \pm b_{-2} \sqrt{b_1} + b_2 \sqrt{b_{-1}})$$

Note that subsystem 1 has boundary solution and subsystem 2 has multiple one. The last has asymptotic

(3.18)
$$z_{1(1,2)} = \pm\sqrt{\epsilon/b_1}, \quad z_{2(1,2)} = (b_{-3} + b_1)\epsilon/b_1 b_3, \quad \epsilon \to 0$$

From relations (2.35), (2.36) we have

(3.19)
$$B_1/B_0 = -(S_1 + 2S_2 + 2S_3),$$
$$B_2/B_0 = (1/2)((B_1/B_0)^2 - (P_1 + 4P_2 + 4P_3 - 4f_{12} + 4f_{13} - 8f_{23})),$$

where $S_i = \sum_{j_i=1}^{M_i} u_i^{-1}(.), P_i = \sum_{j_i=1}^{M_i} u_i^{-2}(.), f_{sl} = \sum_{j_s=1}^{M_s} J_l/J_S u_S^2|_{z=z_{j_s}(0)}$. After substitutions we'll get expressions like (3.11), (3.12).[10]

In [19] some limiting cases were considered ($b_{\pm i} \to \infty$). We consider here an asymptotic behaviour of the reaction rate when b_3 tends to infinity. The kinetic polynomial (3.9) has in this case the form:

(3.20) $\quad (4(b_1 - b_{-1})w^2 + b_2(4b_{-1} - b_2)w - b_{-1}b_2^2)(b_1(b_{-2} + 2w)^2 - (b_{-2} + b_2)^2 w) = 0$

One can show (see [19]) that in the region

(3.21) $\qquad \left\{ \begin{array}{l} b_2 > b_{-2} \\ b_2/2 \le b_1 \le (b_2 + b_{-2})^2/8b_{-2} \end{array} \right.$

multiple physical steady states exist, while in the other regions a single steady state exist. From (3.20) the equations for multiple reaction rates are:

(3.22)

$$w = \left\{ \begin{array}{ll} w_{(1)} = \frac{b_{-2}}{2}(\sqrt{H} - \sqrt{H-1})^2, & \text{if } b_1 < \frac{b_2}{2}(b_2 \le b_{-2}) \text{ or } H \ge 1(b_2 > b_{-2}) \\ w_{(2)} = \frac{b_{-2}}{2}(\sqrt{H} + \sqrt{H-1})^2, & \text{in region (3.21)} \\ w_{(3)} = \frac{b_2}{2}\frac{b_2+\lambda}{4b_1-b_2+\lambda}, & \text{if } b_1 \ge \frac{b_2}{2} \end{array} \right.$$

where $H = \frac{(b_2+b_{-2})^2}{8b_{-2}b_1}$, and $\lambda = \sqrt{b_2^2 + 8b_{-1}(2b_1 - b_2)}$. Steady-state concentration of AZ corresponding reaction rates $w_{(1)}$ and $w_{(2)}$ is low ($\sim b_3^{-1}$) while the concentration of BZ is low at steady state $w_{(3)}$. The branches $w_{(1)}, w_{(2)}, w_{(3)}$ in the space $[b_1, b_2, w]$ form a cusp-surface [30]. The difference between a cuspoid when $b_3 \to \infty$ and the case of finite b_3 is that one line of fold is straight (see [19]). The cross sections of steady-state rate surface, corresponding the dependences of reaction rate on reactant concentrations, have breaks. The break is singular point (double point, point of self-intersection). Consider, for example, kinetic polynomial (3.9)–(3.14) as a function of parameters $\delta = b_3^{-1}, b_1, w : R(w, \delta, b_1) = 0$. When $b_3 \to \infty$ we have $\delta = 0$ and at $b_1 = w = b_2/2$ the following relations are satisfied: $R = R'_w = R'_{b_1} = 0$ and $R''^2_{wb_1} - R''_{ww}R''_{b_1 b_1} > 0$. Thus it's a double point by definition. Therefore the breaks on kinetic curves result from the high rate constant of the reaction between the adsorbed substances. Using the model corresponding to mechanism (3.6) in assumption $b_3 \to \infty$ it has been possible to account for kinetic curve breaks in the oxidation of hydrogen on Pt(111) and Pt(110) [31] as well as on Pd, Ir and on Pd-MOS system [27].

Thus, on the basis of a kinetic polynomial one can obtain simple relations for the limiting cases of kinetic behaviour that describe the specific features of mechanism (3.6). These relations can be used for a preliminary analysis of the experiment and estimation of the parameters in a kinetic model [27, 31]. Finally, a kinetic polynomial permits qualitative classification of the specific kinetic characteristics of reaction mechanism. Most of these features have been studied only numerically (see [29]).

---

[10] To overcome the difficulties when multiple solutions of the equilibrium subsystems are exist the relations (3.18) have been applied.

4. Classical approximations. There are two classical limiting cases where explicit solutions of quasi-steady state equations (2.3) exist, namely, limitation of a step (I) and neighbourhood of equilibrium (II).

I. Existence of rate-determining kth step in the mechanism (2.1) implies that the values of the weights of the kth and $-k$th reactions are small compared to those of the rest of the reactions. In this case one can expand the solution of system (2.3) on the degrees of small parameter b_k (or b_{-k}):

$$(4.1) \qquad w = w_k^{(0)} \left(1 + \sum_{\substack{j=1 \\ j \neq k}}^{n} (-1)^{j+k-1} \frac{\nu_j J_j(\mathbf{z}_0)}{\nu_k J_k(\mathbf{z}_0)} + o(b_k^2) \right),$$

where $\mathbf{z}_0$ is the physical solution of kth equilibrium subsystem. First member of the series (4.1) is the rate of limiting step, calculated at values of intermediate concentrations corresponding the equilibrium of fast steps, i.e. $w_k^{(0)} = (1/\nu_k)u_k(\mathbf{z}_0)$. Exactly the same approximation had been applied usually for the reaction rate calculation under limitation (see [2]). It follows from (4.1) that necessary condition of its applicability is

$$(4.2) \qquad \left| \sum_{\substack{j=1 \\ j \neq k}}^{n} (-1)^{j+k-1} \frac{\nu_j J_j(\mathbf{z}_0)}{\nu_k J_k(\mathbf{z}_0)} \right| \ll 1$$

We can obtain the expression for $w_k^{(0)}$ from the results of section 2. Suppose that 1st step is rate-determining. Expressions for $\mathbf{z}_0$ can be obtained from (2.13) and (2.3.n+1) after substitution $m_j = 0$. Then from (2.3.1) we get

$$(4.3) \qquad w_1^{(0)} = \frac{b_1}{\nu_1} \frac{\prod_{j=2}^{n}(1/\kappa_j)^{\sum_{i=1}^{n-1} \alpha_{1i} A_{ji}/\Delta_1} \left(1 - \prod_{j=1}^{n}(1/\kappa_j)^{\nu_j/\nu_1} \right)}{\left(1 + \sum_{i=1}^{n-1} \prod_{j=2}^{n}(1/\kappa_j)^{A_{ji}/\Delta_1} \right)^{p_1}},$$

where $\kappa_j = b_j/b_{-j}, p_1 = \sum_{j=1}^{n} \alpha_{1j} = \sum_{j=1}^{n} \beta_{1j}$. After some transformations we can obtain the symmetrical form of equation (4.3) for the case when kth step is rate-determining

$$(4.4) \qquad w_k^{(0)} = \frac{b_k}{\nu_k} \frac{\left(1 - \prod_{j=1}^{n}(1/\kappa_j)^{\nu_j/\nu_k} \right)}{\left(\sum_{i=1}^{n} \prod_{j \neq k}^{n}(1/\kappa_j)^{\Delta_{ji}^{'(k)}/p_k N_k} \right)^{p_k}},$$

where $\Delta_{ji}^{'(k)}$ are the cofactors for the corresponding elements of the matrix $\gamma = (\alpha_{ij} - \beta_{ij})$, wherein the kth row is substituted by the row $(\alpha_{k1}, \ldots, \alpha_{kn})$, and N_k is the cofactor for the arbitrary element of the kth row in matrix γ.

Numerator of eq. (4.4) includes the factor

$$\left(1 - \prod_{j=1}^{n}(1/\kappa_j)^{\nu_j/\nu_k} \right) = 1 - (f_-(\mathbf{c})/(Kf_+(\mathbf{c})))^{1/\nu_k}.$$

The rest of eq. (4.4) is the rate of the forward reaction w_+. Thus eq. (4.4) corresponds to expression (1.2) from Boreskov–Horiuti theory. It is of interest that the exponent $1/\nu_k$ in the numerator of (4.4) is the apparent molecularity according to Boreskov and in the denominator (p_k) it is the molecularity generally accepted in kinetics (the number of molecules of intermediates taking part in the rate-determining step). Note that in the general case eq. (4.4) is not a power function of the concentrations for the observed substances, as has been suggested for the derivation of eq. (1.2) in [1,4][11]. The new fact is that it is possible to obtain for a single-route mechanism (2.1) both explicit kinetic equation (4.4) and explicit condition of its applicability (4.2). Expression for z_0 in (4.2) is

(4.5)
$$z_{l0} = \frac{\prod_{j\neq k}^{n}(1/\kappa_j)^{\Delta_{ji}^{'(k)}/p_k N_k}}{\sum_{i=1}^{n}\prod_{j\neq k}^{n}(1/\kappa_j)^{\Delta_{ji}^{'(k)}/p_k N_k}},$$

From (4.4) and (4.5) one can obtain the expressions for the observed reaction order n_μ with respect to the substance C_μ, and the observed activation energy E_0:

(4.6)
$$n_\mu = \frac{\partial \ln |w|}{\partial \ln c_\mu} = \gamma'_{k\mu} + N_k^{-1}\sum_{i=1}^{n} z_{i0}\left(\sum_{j\neq k}^{n}\Delta_{ji}^{'(k)}(\gamma'_{j\mu} - \gamma''_{j\mu})\right) + \frac{\Gamma_{\mu I} - \Gamma_{\mu T}}{\nu_k\left(\prod_{j=1}^{n}\kappa_j^{\nu_j/\nu_k} - 1\right)},$$

(4.7)
$$E_0 = \frac{\partial \ln |w|}{\partial(-1/RT)} = E_k - N_k^{-1}\sum_{i=1}^{n} z_{i0}\left(\sum_{j\neq k}^{n}\Delta_{ji}^{'(k)}Q_j\right) - \frac{Q}{\nu_k\left(\prod_{j=1}^{n}\kappa_j^{\nu_j/\nu_k} - 1\right)},$$

where E_k and E_{-k} are activation energies of rate-determining step reactions (forward and back), $Q_j = E_{-j} - E_j$ is the heat of jth reaction, $Q = \sum_{i=1}^{n}\nu_i Q_i$ is the heat of net reaction, $\Gamma_{\mu I(T)} = \sum_{s=1}^{n}\gamma_{s\mu}^{'('')}\nu_s$ are the stoichiometric coefficients for the μth substance in the net reaction equation.

Equations (4.6) and (4.7) can be applied to obtain the relation between n_μ^+ and n_μ^- that are observed reaction orders with respect to the substance C_μ, and the observed activation energies E_0^+ and E_0^- far from equilibrium. Assuming that $n_\mu^\pm$ and $E_0^\pm$ are determined for the same composition of the surface (i.e. at the same κ_j), we obtain

(4.8)
$$\frac{n_\mu^+ - n_\mu^-}{\Gamma_{\mu I} - \Gamma_{\mu T}} = \frac{E_0^- - E_0^+}{Q} = \frac{1}{\nu_k} = M_k$$

Formulae (4.8) are similar to those given by Boreskov [4,8], but for their application it is not necessary that the kinetic equation should be exponential. They can be used to determine a stoichiometric number for the rate-determining step according to the experimental data. It follows from (4.6), (4.7) that observed order and

---

[11] We must note that attempt [32] to obtain the power-form Boreskov–Horiuti equation (1.2) seems no correct since these authors do not consider the balance equations like (2.3.n+1).

observed activation energy include three components: limiting step parameter (1), addendum accounted for detailed reaction mechanism (b), and addendum described reversibility of the overall reaction (c). For linear mechanism it's possible to obtain from (4.6) and (4.7) structured formulae like in [14].

Consider, for example, the mechanism (3.6) when second step is the rate-determining. We have $N_2 = \nu_2 = 2, p_2 = 1$,

$$
\gamma'^{(2)} = \begin{pmatrix} 2 & -2 & 0 \\ 1 & 0 & 0 \\ -2 & 1 & 1 \end{pmatrix},
$$

$\Delta_{11}'^{(2)} = \begin{vmatrix} 0 & 0 \\ 1 & 1 \end{vmatrix} = 0, \Delta_{31}'^{(2)} = \begin{vmatrix} -2 & 0 \\ 0 & 0 \end{vmatrix} = 0, \Delta_{12}'^{(2)} = -\begin{vmatrix} 1 & 0 \\ -2 & 1 \end{vmatrix} = -1, \Delta_{32}'^{(2)} =$

$-\begin{vmatrix} 2 & 0 \\ 1 & 0 \end{vmatrix} = 0, \Delta_{13}'^{(2)} = \begin{vmatrix} 1 & 0 \\ -2 & 1 \end{vmatrix} = 1, \Delta_{33}'^{(2)} = \begin{vmatrix} 2 & -2 \\ 1 & 0 \end{vmatrix} = 2.$ Then, by (4.1), (4.4)

and (4.5) we get

$$
w = w_2^{(0)} \left(1 - \frac{b_2 b_3 (y_0 - x_0) + b_{-2}(2b_{-3}z_0 + b_3 y_0) + 2(2b_1 b_{-2} z_0 + 2b_{-1}(b_2 + b_{-2})x_0)}{4(b_1 b_3 z_0 (x_0 - y_0) + b_{-1}x_0(2b_{-3}z_0 + b_3 x_0))} + o(b_2^2) \right),
$$

where

$$
w_2^{(0)} = \frac{b_{-2}}{2} \frac{\kappa_2 \kappa_3 \sqrt{\kappa_1} - 1}{\kappa_3 \sqrt{\kappa_1} + \kappa_1 \kappa_3 + 1},
$$

$$
\frac{x_0}{\kappa_1 \kappa_3} = y_0 = \frac{z_0}{\kappa_3 \sqrt{\kappa_1}} = \frac{1}{\kappa_3 \sqrt{\kappa_1} + \kappa_1 \kappa_3 + 1}.
$$

From expression (4.6)

$$
n_{A_2} = \frac{1}{2} \left(y_0 - x_0 + \frac{1}{\sqrt{K c_{A_2} c_B^2 / c_{AB}^2} - 1} \right).
$$

When c_B is large $n_{A_2} = n_{A_2}^+ = (1/2)(y_0 - x_0)$, when c_B is small $n_{A_2} = n_{A_2}^- = (1/2)(y_0 - x_0 - 1)$. Thus $n_{A_2}^+ - n_{A_2}^- = 1/2$ in accordance with (4.8).

II. In the neighbourhood of equilibrium the condition $K \simeq f_-(\mathbf{c})/f_+(\mathbf{c})$ is fulfilled. Then the value of C in eq. (2.42) is small (i.e. $C = \epsilon \to 0$). Consider the possibilities to find the solution of eq. (2.42) in the form of ϵ-power series.

(A) If one express the parameter, say b_1, as a function of ϵ and substitute it in the coefficients of (2.42) then last takes the form

(4.9) $\qquad B_l = B_{leq} + \epsilon P_l(\epsilon, \mathbf{b}), \quad B_s' = B_{seq}' + \epsilon P_s'(\epsilon, \mathbf{b}),$

where subscript "eq" corresponds the equilibrium value. It follows from Newton diagram [33] of equation (2.42) with coefficients (4.9) that there are two classes of solutions when $\epsilon \to 0$:

(1) ϵ-order solutions, corresponding the descending part of diagram

(4.10) $\qquad w_s = v_{1s}\epsilon, \quad s = 1, \ldots, p$

(2) Solutions of order $o(1)$ corresponding the horizontal part of diagram

$$(4.11) \qquad\qquad w_l = v_{0l}, \quad l = p+1,\ldots,L$$

The values of v_{1s} and v_{0l} are the roots of polynomials

$$(4.12) \qquad B_{peq}v_1^p + B'_{(p-1)eq}v_1^{p-1} + \cdots + B'_{1eq}v_1 + B'_{0eq} = 0,$$

$$(4.13) \qquad B_{Leq}v_0^{L-p} + \cdots + B_{peq} = 0.$$

Reaction rate near equilibrium is one of solutions (4.10). Other solutions are non-physical.

(B) Suppose that $L > p$. After substitution $w = \epsilon v$ into (2.42) and introduction of parameter $t \in [0,1]$ we'll get

$$(4.14) \quad (B_L\epsilon^{L-p-1}v^L + \cdots + B_{p+1}v^{p+1})(\epsilon t) + B_p v^p + B'_{p-1}v^{p-1} + \cdots + B'_0 = 0.$$

Then the solution of (4.14) at $t = 1$ can be obtained in the form

$$(4.15) \qquad\qquad w = v^*\epsilon + \left.\frac{dv}{d\tau}\right|_{\tau=0, v=v^*} \epsilon^2 + \ldots,$$

where $\tau = \epsilon l$ and v^* is the root of equation $B_p v^p + \cdots + B'_0 = 0$.

Approximations (A) and (B) give the same order error, but examples show that approximation (4.15) can be more precise.

Thus, the first member of steady-state reaction rate approximation in the neighbourhood of equilibrium is the root of polynomial corresponding the first $p + 1$ coefficients of the kinetic polynomial (2.42). One can obtain the explicit form of kinetic equation near equilibrium. Let $p = 1$. From (4.10) we have

$$(4.16) \qquad\qquad w = -B'_{0eq}\left(\prod_{i=1}^{n} b_i^{\nu_i} - \prod_{i=1}^{n} b_{-i}^{\nu_i}\right) / B_{1eq}$$

On the other hand, in accordance with theorem 2.2

$$B_1/B_0 = -\sum_{i=1}^{n}\nu_i S_i,$$

where $S_i = \sum_{j_i=1}^{M_i} 1/u_i(z_{j_i}(0))$. Consider one of S_i, for instance S_1. It follows from the proof of theorem 2.1 that

$$(4.17) \qquad\qquad S_1 = Y / \left(\prod_{i=1}^{n}(b_i/b_{-1})^{\nu_i} - 1\right),$$

where

$$(4.18) \qquad Y = \sum_{j_i=1}^{M_1} \prod_{s \neq j_1}^{M_1} u_1(\cdot) / \left(\prod_{j_i=1}^{M_{1b}} u_1(\cdot) \prod_{fin}^{M_{1in}} u_{1-}(\cdot)(-1)^{\nu_1 - 1}\right).$$

Here the subscript "b" corresponds the boundary solutions of the equilibrium subsystem, $u_{1-}(\cdot)$ is the rate of first step in back direction. The value of M_{1in} is $|\nu_1|$. We have to define Y value in the equilibrium point (i.e. $\prod_{i=1}^{n}(b_i/b_{-i})^{\nu_i} = 1$). Then the numerator of Y will contain only one addendum, corresponding the product on all solutions of equilibrium subsystem beside physical one (subscripted as "ph"):

$$(4.19) \qquad Y = (-1)^{\nu_1-1} u_{-1eq}^{-1} \prod_{j_{in}\neq j_{ph}}^{\nu_1} (u_{1+}(\cdot)/u_{1-}(\cdot) - 1),$$

where $u_{-1eq} = u_{+1eq} = u_{1eq}$ is equilibrium exchange rate of the first step and $u_{1+}(\cdot)$ is the rate of the first step in forward direction. It follows from (2.21) that

$$(4.20) \qquad \prod_{j_{in}\neq j_{ph}}^{\nu_1} (u_{1+}(\cdot)/u_{1-}(\cdot) - 1) = \prod_{l=1}^{\nu_1-1} (\exp(2\pi\mathfrak{F}l/\nu_1) - 1) \equiv (-1)^{\nu_1-1}\nu_1$$

Then from relation (4.19) $Y = \nu_1/u_{1eq}$. The result is exactly the same for other S_i. From eq (2.35) in the neighbourhood of equilibrium we have

$$(4.21) \qquad 1/w = \left(\sum_{k=1}^{n} \nu_k^2/u_{keq}\right) \Big/ \left(\prod_{i=1}^{n}(b_i/b_{-i})^{\nu_i} - 1\right)$$

This relation is valid at appropriate p values [27].

After substitution of the expression (4.5) into eq. (4.21) we'll get

$$(4.22) \qquad w = \frac{\prod_{j=1}^{n}\kappa_j^{\nu_j} - 1}{\sum_{k=1}^{n}\nu_k^2 b_k^{-1}\left(\sum_{i=1}^{n}\prod_{j\neq k}^{n}(1/\kappa_j)^{\Delta_{ji}'^{(k)}/p_k N_k}\right)^{p_k}}\Bigg|_{eq},$$

where the subscript "eq" means that the denominator in eq. (4.22) is calculated at the equilibrium point. The numerator of eq. (4.22) is equal to $Kf_+(\mathbf{c})/f_-(\mathbf{c}) - 1 \simeq -\Delta G/(RT)$, and the denominator is the inverse of the exchange rate of overall reaction

$$1/w_{eq} = \sum_{k=1}^{n} \nu_k^2/u_{keq}$$

and, hence, eq. (4.22) corresponds to equation (1.3). The value $1/w_{eq}$ is proportional to the characteristic time of the quasi-steady-state process of the reaction in the closed system. Thus kinetic equation (4.22) includes values determined in various experiments: steady-state (w), relaxation (w_{eq}) and isotopic (u_{keq}). It is essential that for these values explicit equations have been obtained through the parameters of reaction mechanism (2.1).

Near equilibrium the observed values n_μ and E_0 take the form

$$n_\mu = \frac{\Gamma_{\mu I} - \Gamma_{\mu T}}{-\Delta G/(RT)}$$

$$E_0 = \frac{Q}{\Delta G/(RT)}$$

and are independent of the detailed reaction mechanism and controlled only by the thermodynamic characteristics of the overall reaction.

A case is possible when the step limitation takes place in the neighbourhood of equilibrium. Here eqs. (4.4) and (4.22) lead to the same result

$$(4.23) \qquad w = \frac{\nu_k^{-2} b_k \left(\prod_{j=1}^{n} \kappa_j^{\nu_j} - 1 \right)}{\left(\sum_{i=1}^{n} \prod_{j \neq k}^{n} (1/\kappa_j)^{\Delta_{ji}'^{(k)}/p_k N_k} \right)^{p_k}} \Bigg|_{eq}.$$

The equations (4.4), (4.22), (4.23) can be applied for different purposes: to construct a kinetic model on the basis of the detailed reaction mechanism (a), and to verify the hypothesis concerning the reaction mechanism (b).

5. Applications. It was shown in section 3 that kinetic polynomial can be used for asymptotic analysis of steady-state multiplicity. On the other hand, it can be used for direct computation of the all steady states. After variable elimination the problem reduced to the finding of the roots of single-variable polynomial (see [34]). Here we'll consider the application of kinetic polynomial for description the steady-state dependences of the reaction rate (I) and the study of the parameter identifiability of given kinetic model (II).

I. The kinetic polynomial can be written as

$$(5.1) \qquad \sum_{j=1}^{m} f_j(\mathbf{k}) \phi_j(\mathbf{c}, w) = 0, \quad \phi_j(\mathbf{c}, w) = c_1^{\xi_{j1}} \ldots c_l^{\xi_{jl}} w^{\xi_j,(l+1)},$$

where c_i are the concentrations of the species observed, $f_j(\mathbf{k})$ are the polynomials in the reaction rate constants $\mathbf{k}$, m is the number of different monomials ϕ_j, and ξ_{ji} are integers. The problem of finding implicit kinetic dependences may have either one of two formulations: the reaction mechanism and the corresponding kinetic polynomial in the form (5.1) are specified, and the polynomial coefficients $f_j = f_j(\mathbf{k})$ are to be determined (A); the reaction mechanism is unknown, and the kinetic dependence in the form of polynomial (5.1) and its parameters are to be found (B).

In terms of eq. (5.1) the problem (A) formulated as follows: the experimental data c_i and w_i being known, the f_j values optimal in terms of a certain criterion are to be found. Usually the optimization criterion has the form

$$(5.2) \qquad \sum_{i=1}^{N} (w_i^e - w_i)^2 g_i \rightarrow \min_k,$$

where w_i^e and w_i are the steady-state reaction rates measured experimentally and calculated by eq. (5.1), g_i are the weights of the experimental points, and N is the number of measurements.

The kinetic polynomial (5.1) admits various approximations of criterion (5.2). The lhs of eq. (5.1) may be expanded in a series[12] in the vicinity of w_i^e

$$(5.3) \quad \sum_{j=1}^{m} f_j(\mathbf{k}) \phi_j(\mathbf{c}_i, w_i) = \sum_{j=1}^{m} f_j(\mathbf{k}) \phi_j(\mathbf{c}_i, w_i^e) + \sum_{j=1}^{m} f_j(\mathbf{k}) \frac{\partial \phi_j}{\partial w} \Bigg|_{w=w_i^e} (w_i - w_i^e) + \ldots$$

---

[12] $c_1, \ldots, c_l$ are assumed to be measured without experimental error.

By truncating expansion (5.3) one obtains approximations of different orders for criterion (5.2).

The zero order approximation

$$(5.4) \qquad \sum_{j=1}^{m} f_j(\mathbf{k})\phi_j(\mathbf{c}_i, w_i^e) = 0$$

the first order one being

$$(5.5) \qquad \sum_{i=1}^{N} g_i \left(\frac{\sum_{j=1}^{m} f_j(\mathbf{k})\phi_j(\mathbf{c}_i, w_i^e)}{\sum_{j=1}^{m} f_j(\mathbf{k})\partial\phi_j/\partial w|_{w=w_i^e}} \right)^2 \to \min_{k}.$$

Usually the set of equations (5.4) is overdetermined ($N \geq m, N \geq \dim \mathbf{k}$). A simple way to solve the equations is to use the linear least squares method (LSM).

$$(5.6) \qquad \sum_{i=1}^{N} \left(\sum_{j=1}^{m} f_j\phi_j(\mathbf{c}_i, w_i^e) \right)^2 \to \min_{f}, \quad \|\mathbf{f}\| = const .$$

As an example consider mechanism (3.6). The experimental points are the values of function $w^e(c_{A_2})$ calculated at the following parameters: $b_{-3} = 0.5, b_{-2} = 1.0, b_{-1} = 0.1, k_1 = 1.0, b_2 = 2.0, b_3 = 10^4, 0 < c_{A_2} < 1.3$. Note that the "experimental" points were selected in a region with non-unique steady states. The parameters of the kinetic polynomial (3.9)–(3.14) corresponding to mechanism (3.6) are as follows (see eq. (5.1)): $l = 1, m = 13, (\xi)^T = \begin{pmatrix} 0 & 1 & 0 & 1 & 0 & 1 & 2 & 0 & 1 & 2 & 0 & 1 & 2 \\ 0 & 0 & 1 & 1 & 2 & 2 & 2 & 3 & 3 & 3 & 4 & 4 & 4 \end{pmatrix}$.
The formulas for the coefficients $f_j(\mathbf{k})$ are given in [21].

Table 5.1 lists the values of $x_j = f_j/f_{13}$ determined by solving the set of normal equations by the Gauss method, (the number of points is 13, the number of points in the multiple steady-state region is 10) along with the true values of x_j^t. As is seen from Table 5.1, all the parameters except x_1 and x_{11} were estimated satisfactorily (x_1 and x_{11} may be taken to be equal to zero since the backward reaction 3 in mechanism (3.6) may be disregarded for the given set of rate constants). The computations proved the accuracy of determination of $\mathbf{x}$ to depend on the number and character of experimental points, it deteriorates with an increase in the density of the points, because problem (5.6) is ill-conditioned. The problem (5.6) was also solved using the algorithm of the singular value decomposition SVD [35] which reduces the general LSM problem to the one with diagonal matrix. The problem conditionality may be improved by introducing a bound for the minimal singular value. The calculation results $X_{(1)}$ for such an example together with the singular values σ are presented in Table 5.1 (the number of points is 50, the number of points in the multiple steady-state region is 36). The estimates $X_{(1)}$ correspond to the choice of the bound of the minimal singular value which exceeds σ_{12}. As follows from Table 5.1, $\sigma_1/\sigma_{12} \sim 10^{10}$, i.e. the problem is very ill-conditioned. Estimates $X_{(1)}$ are worse then x_j, nevertheless their signs and orders are the same as for the true values, except for x_1 and x_{11}.

Estimates obtained by using the SVD method may be improved with the aid of ridge regression [36]. Example of its implementation is given in [21]. Thus:

(1) the zero order approximation produces quite adequate estimates of x_j in some cases;

(2) despite the fact that problem (5.6) is ill-conditioned it may be regularized by standard methods of the linear LSM;

(3) there exist a possibility of finding quickly an initial approximation for estimates of the kinetic parameters by linear methods.

TABLE 5.1

Solution of the problem (5.6) for the reaction scheme (3.6) and different samples

x_j^t	i	x_j	$X_{(1)}$	σ
$1.5625E - 12$	1	$-1.2589E - 08$	$2.989E - 09$	$6.735E\ 06$
$-2.5000E - 02$	2	$-2.5023E - 02$	$-4.094E - 02$	$3.772E\ 04$
$2.2500E - 01$	3	$2.2521E - 01$	$3.685E - 01$	$1.845E\ 02$
$-2.9996E - 01$	4	$-3.0016E - 01$	$-2.887E - 01$	$1.131E\ 01$
$1.7999E\ 00$	5	$1.8006E\ 00$	$1.125E\ 00$	$3.218E\ 00$
$-9.2480E - 01$	6	$-9.2506E - 01$	$-8.288E - 01$	$6.464E - 01$
$2.5005E - 01$	7	$2.5012E - 01$	$2.902E - 01$	$2.752E - 01$
$2.2497E - 01$	8	$2.2630E - 01$	$1.457E\ 00$	$1.482E - 01$
$-3.1499E\ 00$	9	$-3.1536E\ 00$	$-3.611E\ 00$	$1.394E - 02$
$1.0001E\ 00$	10	$1.0014E\ 00$	$1.027E\ 00$	$6.821E - 03$
$2.5000E - 11$	11	$-9.0114E - 04$	$-5.473E - 01$	$1.510E - 03$
$-9.9990E - 02$	12	$-9.8605E - 02$	$-4.677E - 02$	$3.222E - 04$
1.0	13	1.0	1.0	$--$

Formally, the problem (B) reduces to the finding the form of the monomials $\phi_j(\mathbf{c}, w)$ and coefficients f_j of eq. (5.1) which fit best a certain criterion used for the description of experiment. In the numerical experiments presented in [21] this problem was solved by a step-ordering method [37] which permits one to construct polynomials with a maximal convergence speed. The informative population of monomials was estimated with the aid of the slip-control. The results presented in [21] show that in spite of not all of the monomials of the original polynomial appear in the reconstructed polynomial, the signs and orders of the recovered coefficients coincide with the true ones. It's important that the deduced polynomial represents qualitatively the kinetic dependence characterized by multiple steady states: it incorporates terms with w^3, w^4.

The method offered here consists essentially in linearization of the initial problem. Linearization is commonly used to evaluate kinetic parameters, e.g. Arrhenius equation parameters and reaction orders; in the case of explicit rational kinetic dependences form like (5.1) has been used, for instance in [38,39]. Transformation of a kinetic model to form (5.1) offers the following advantages:

(1) It permits solving the problem rapidly by the multiple linear regression method;

(2) When the linear problem is nonconfluent (or may be regularized) the solution uniqueness is guaranteed;

(3) The results of solution of problem (5.6) may be utilized to find the initial approximation when estimating the original kinetic parameters by nonlinear procedure;

(4) When eq. (5.1) is used experimental data characterized by multiple steady states are treated in the same way as those with a single steady state;

(5) Form (5.1) enables one to formulate the problem of determining steady-state relations in the absence of information on the reaction mechanism.

II. The inverse chemical kinetics problem does not necessarily have a unique solution, in particular it may posses a continuum of solutions. When analyzing this problem one should take into account possible relations between kinetic parameters. The techniques for determining these relations was suggested in [40]. However this technique can involve rather cumbersome calculations, first of all for nonlinear mechanisms. Criteria for the appearance of relations between parameters stemming from the structure of the reaction graph were presented in [41] for linear mechanism. On the basis of kinetic polynomial concept we can suggest a new approach to the determination of relations between parameters of nonlinear kinetic models.

Under assumption of quasi-steady-state approximation kinetic model can be reduced after elimination of unknown intermediate concentrations to the form

$$(5.7.1) \qquad \frac{d\mathbf{c}}{dt} = G\mathbf{w},$$

$$(5.7.2) \qquad \sum_{j_i=1}^{m_i} f_{ij_i}(\mathbf{k})\phi_{ij_i}(\mathbf{c}, w_i) = 0, \quad i = 1, \ldots, P,$$

where $\mathbf{w}$ are either the rates on reaction routes or their linear combinations; $P = \dim \mathbf{w} \leq R$ (R is the number of reaction routes); G is an integer-valued matrix; and $\mathbf{k}$ is the vector of kinetic constants[13].

Therefore if it is possible to replace variables $\mathbf{k} \to \rho$ in (5.7.2) so that $\dim \rho = r < \dim \mathbf{k} = n$, then system (5.7) also depends only on ρ, i.e. singularity takes place.

---

[13]It is assumed that expressions (5.7.2) are normalized to one of $f_{ij_i}(\mathbf{k})$. Thus m_i refers to the number of independent coefficients of (5.7.2).

It was proved [20] that relations between constants $\mathbf{k}$ can follow from matrix

$$(5.8) \qquad B(m \times n) = \begin{pmatrix} \partial f_{11}/\partial k_1 & \cdots & \partial f_{11}/\partial k_n \\ \vdots & \cdots & \vdots \\ \partial f_{1m_1}/\partial k_1 & \cdots & \partial f_{1m_1}/\partial k_n \\ \vdots & \cdots & \vdots \\ \partial f_{P1}/\partial k_1 & \cdots & \partial f_{P1}/\partial k_n \\ \vdots & \cdots & \vdots \\ \partial f_{Pm_p}/\partial k_1 & \cdots & \partial f_{Pm_p}/\partial k_n \end{pmatrix},$$

where $m = \sum_{i=1}^{P} m_i$. There are at least two situations which may cause the appearance of such relations, namely, when the number m of coefficients in forms (5.7.2) is less than the number n of rate constants and when $m \geq n$ and $rank\ (B) < n$, i.e. partial derivatives of the functions f are linearly dependent. The former situation does not exclude relations arising due to the latter cause (see example below).

Thus the procedure of determining the relations and performing a transformation to new parameters amounts for system (5.7) to the following: reduction of the system of quasi-steady equations to a single polynomial (or, in the case of multiroute mechanism, to several polynomials) in $\mathbf{w}$, transformation of the resulting equations to form (5.7.2), determination of the rank of Jacobian (5.8), and determination of the relations between the rate constants.

The last can be found in the following way:

(1) r linearly independent functions f from form (5.7.2) are taken as coordinates ρ;

(2) formulae for the remaining functions f are written;

(3) the parameters $\mathbf{k}$ are eliminated from resulting algebraic system of equations.

Now we indicate some properties of the algorithm under discussion in which it differs from the procedure suggested in [40]: explicit functions of the rate constants $\mathbf{k}$ are analysed and therefore relations between them (their functional combinations) can be found directly without investigation the Jacobian matrix; the factors causing the appearance of relations between kinetic parameters can be revealed; the algorithm requires computer calculation of the rank of a matrix whose elements are rational functions. To calculate the rank of such a matrix use can be made of the effective analytical computer methods.

Consider for example a liquid-phase hydrocarbon oxidation scheme (a fragment)

$$(5.9) \qquad \begin{aligned} RH + O_2 &\xrightarrow{k_0} R + HO_2, \quad R + R \xrightarrow{k_4}, \\ R + O_2 &\xrightarrow{k_1} RO_2, \quad RO_2 + R \xrightarrow{k_5}, \\ RO_2 + RH &\xrightarrow{k_2} ROOH + R, \quad RO_2 + RO_2 \xrightarrow{k_6}, \\ ROOH &\xrightarrow{k_3} RO + OH \end{aligned}$$

Here RH is a hydrocarbon; R, RO_2, RO, and HO_2 are radicals, and $ROOH$ is the observed species. It is assumed that $[RH]$ and $[O_2]$ are constant, and quasi-steady-state conditions hold for $[R]$ and $[RO_2]$. Mechanism (5.13) is not single-routed but it is possible to derive a polynomial with respect to the rate of formation of the observed species $ROOH$ from the quasi-steady-state conditions for the radicals R and RO_2.

Let $P = [ROOH], w_0 = k_0[O_2]$ (the value of k_0 is known), $\rho_1 = k_1[O_2]/w_0$, $\rho_2 = k_1[RH]/w_0, \rho_3 = k_3/w_0, \rho_4 = 2k_4/w_0, \rho_5 = k_5/w_0, \rho_6 = 2k_6/w_0, \theta = tw_0, u = \rho_2[RO_2]$. Then we have

(5.10.1) $$dP/d\theta = u - \rho_3 P,$$

(5.10.2) $$u^4 + au^3 + bu^2 + cu + d = 0,$$

where

$$a = 2\rho_2^2/\rho_6, \quad b = \rho_2^2(\rho_1^2\rho_6 + \rho_2^2\rho_4 + 2\rho_1\rho_2\rho_5 - \rho_5^2)/(\rho_6(\rho_4\rho_6 - \rho_5^2)),$$
$$c = 2\rho_1\rho_2^3\rho_5/(\rho_6(\rho_4\rho_6 - \rho_5^2)), \quad d = -\rho_1^2\rho_2^4/(\rho_6(\rho_4\rho_6 - \rho_5^2)).$$

Expression (5.10.2) is a polynomial of the fourth order and hence the number of the variables on which its solution depends is less than the number of rate constants in mechanism (5.9). Therefore even formula (5.10.2) alone permits drawing a conclusion about relations between the parameters. To determine the exact number of independent parameters we find the rank of the matrix

(5.11) $$B = \begin{pmatrix} 0 & \partial a/\partial\rho_2 & 0 & 0 & \partial a/\partial\rho_6 \\ \partial b/\partial\rho_1 & \partial b/\partial\rho_2 & \partial b/\partial\rho_4 & \partial b/\partial\rho_5 & \partial b/\partial\rho_6 \\ \partial c/\partial\rho_1 & \partial c/\partial\rho_2 & \partial c/\partial\rho_4 & \partial c/\partial\rho_5 & \partial c/\partial\rho_6 \\ \partial d/\partial\rho_1 & \partial d/\partial\rho_2 & \partial d/\partial\rho_4 & \partial d/\partial\rho_5 & \partial d/\partial\rho_6 \end{pmatrix}$$

The diagonal minor of the third order of B is not identically zero. At the same time the two minors of the fourth order bordering it are identically zero and hence the rank of matrix (5.11) is three. Thus, taking into account additionally the parameter ρ_3, we have 4 independent kinetic parameters.

The simplest system of functional combinations is

$$\rho_1' = \rho_2^2/\rho_6, \quad \rho_2' = \rho_1\rho_2/\rho_5, \quad \rho_3' = \rho_3, \quad \rho_4' = \rho_4\rho_6/\rho_5^2$$

and

$$a = 2\rho_1', \quad b = (\rho_1')^2 + ((\rho_1' + \rho_2')^2 - \rho_1')/(\rho_4' - 1),$$
$$c = 2\rho_1'\rho_2'/(\rho_4' - 1), \quad d = -\rho_1'(\rho_2')^2/(\rho_4' - 1)$$

Note that functional combinations can also be found immediately from the form of a, b, c, and d.

Thus the reduction of the system of quasi-steady-state equations to polynomials in the rates of formation of the species observed and the investigation of the coefficients of these polynomials permits determination of relations between the rate constants of nonlinear kinetic models. The determination of the relations between the rate constants must precede the solution of any inverse kinetic problem. To determine the relations, use is made of information about the reaction mechanism and the type of experiment (i.e. what parameters are measured and which of the species observed change and which are maintained constant). In doing this, no numerical values of the experimental data are employed.

6. Algorithm of computation. Formulas for coefficients of kinetic polynomial as functions of the equilibrium subsystems roots have been obtained in section 2 (see eqs. (2.35) and (2.36)). These expressions are useful for theoretical analysis and obtaining of chemical corollaries, but non-effective for calculation. New expressions for kinetic polynomial coefficients have been obtained recently [42].

Let $u_j(\mathbf{z}) = b_j \prod_{i=1}^{n} z_i^{\alpha_{ji}} - b_{-j} \prod_{i=1}^{n} z_i^{\beta_{ji}}$ (for notation see section 2) and $u_{n+1} = z_1 + \cdots + z_n - 1$. It follows from the assumptions of section 2 (there are no roots of the system (2.3) at $w = 0$) and Hilbert theorem about zeros [16] that there exists a set of homogeneous polynomials $a_{jk}(\mathbf{z}), j, k = 1, \ldots, n$ such that

$$(6.1) \qquad z_j^{L+1} = \sum_{k=1}^{n} a_{jk}(\mathbf{z}) u_k(\mathbf{z}), \quad j = 1, \ldots, n$$

The degree L in (6.1) can be chosen $r_1 + \cdots + r_n - n$ or less, where $r_j = \sum_{i=1}^{n} \alpha_{ji} = \sum_{i=1}^{n} \beta_{ji}$ (see (2.2)) is the degree of the $u_j(\mathbf{z})$.

Denote as A the matrix $(a_{jk}(\mathbf{z}))_{j,k=1}^{n}$ and as J_s the Jacobian of sth subsystem. The following facts have been proved in [42]:

THEOREM 6.1. *If the system (2.3) has no roots at $w = 0$ then*

$$(6.2) \qquad \frac{B_1}{B_0} = d_1 = \sum_{s=1}^{n} (-1)^{n+s-1} \nu_s \Xi \left(\frac{(z_1 + \cdots + z_n)^{r_s-1} J_s \det A}{z_1^L \cdots z_n^L} \right),$$

where Ξ is a functional which extracts from Lorain polynomial $Q = P(z_1, \ldots, z_n) / (z_1^{L_1} \ldots z_n^{L_n})$ its free member (i.e. the coefficient corresponding the monomial $z_1^{L_1} \ldots z_n^{L_n}$ in the polynomial $P(z_1, \ldots, z_n)$).

THEOREM 6.2. *When conditions of Theorem 6.1 are fulfilled*
$$(6.3)$$
$$\frac{2B_2}{B_0} - \left(\frac{B_1}{B_0} \right)^2 = d_2 = \sum_{k=1}^{n} \sum_{s,l=1}^{n} (-1)^{n+s-1} \nu_s \nu_l \Xi \left(\frac{(z_1 + \cdots + z_n)^{r_s+r_l-1} J_s \det A a_{kl}}{z_1^L \cdots z_k^{2L+1} \cdots z_n^L} \right),$$

In general case we have

THEOREM 6.3. *When conditions of Theorem 6.1 are fulfilled and $s > 0$*

$$(6.4) \qquad d_s = \sum_{l=1}^{n} \sum_{\substack{\|\alpha\|=s \\ \alpha_1 = \cdots = \alpha_{l-1} = 0, \ \alpha_l > 0}} (-1)^{n+l-1} \frac{s!}{\alpha!} \nu^\alpha \sum_{j_1, \ldots, j_{s-1}=1}^{n} c_{j_1 \ldots j_{s-1}} \Xi(Q(\mathbf{z})),$$

where

$$Q(z) = \frac{(z_1 + \cdots + z_n)^{r_l \alpha_l + \cdots + r_n \alpha_n - 1} a_{j_1 l} \ldots a_{j_{s-1} n} \det A J_l}{z_{j_1}^{L+1} \ldots z_{j_{s-1}}^{L+1} z_1^L \ldots z_n^L},$$

Here $d_s = d^{(s)} \ln R(w)/dw^s$ (see section 2), $\alpha = (\alpha_1, \ldots, \alpha_l, \ldots, \alpha_n)$, $\alpha! = \alpha_1! \ldots \alpha_n!$, $\nu^\alpha = \nu_1^{\alpha_1} \ldots \nu_n^{\alpha_n}$, $c_{j_1} \ldots j_{s-1} = \beta_1! \ldots \beta_n!$, where $\beta_1, \beta_2 \ldots$ are the numbers of "1", "2", ... in the set of subscripts $j_1, \ldots, j_{s-1}$.

Thus the algorithm of kinetic polynomial computation includes the steps:

(1) calculation of matrix A,

(2) calculation of d_s by formula (6.4), and

(3) calculation of B_k/B_0 by formula (2.25)

Possible way to calculate matrix A is the method of undefined coefficients which reduces the problem to the solving of some linear equation system. It was applied when analysing the examples in [42]. Explicit expressions for $a_{jk}(z)$ are given in [43] for the special form systems.

We find however that the method of $a_{jk}(z)$ generation via the Grobner bases procedure [44,45] is more convenient for realisation.

This method is technique that gives the solutions of such problems as finding of exact solutions of algebraic equation systems (i), finding of polynomial solutions of the systems of homogeneous linear equations (ii), and other problems of polynomial ideal theory.

For our purpose (calculation of matrix A) we applied Grobner bases to solve problem (ii)[14]. The algorithm is as follows

(1) Grobner basis calculation for the system $u_1(z), \ldots, u_n(z)$. We find simultaneously a linear representation of Grobner base polynomials g_i through the original polynomials $u_i(z)$:

(6.5)
$$g_i = \sum_{j=1}^{n} u_j X_{ij}, \quad i = 1, \ldots, m,$$

where X_{ij} are polynomials in $z_1, \ldots, z_n$;

(2) Reduction to zero (in sense of [44]) of the monomials $z_j^{L+1} (j = 1, \ldots, n)$ mod $(g_1, \ldots, g_m)$. Simultaneously coefficients of the representation

(6.6)
$$z_k^{L+1} = \sum_{i=1}^{m} h_i^{(k)} g_i$$

are calculated;

(3) Calculation of the $a_{jk}(z)$ by formula

(6.7)
$$a_{jk}(z) = \sum_{i=1}^{m} h_i^{(k)} X_{ij}$$

---

[14]Nevertheless one can obtain kinetic polynomial by solving problem (i), but our experience proved that direct application of Grobner bases to kinetic systems may be difficult in computational aspect.

Second part of the algorithm calculates B_i by formulas (6.4) and (2.25).

The algorithm has been realized in computer algebra system REDUCE [46][15]. The program description is given in [47]. The results of matrix A calculation are presented below for adsorption mechanism (3.6)

$$a_{1,1} = (b_2 b_3^2 b_{-2} z_3)/(b_1 b_2^2 b_3^2 - b_{-1} b_{-2}^2 b_{-3}^2)$$

$$a_{1,2} = (z_2 b_2 b_3 b_{-1} b_{-2} b_{-3} z_1 + z_2 b_3 b_{-1} b_{-2}^2 b_{-3} z_3 + b_1 b_2^2 b_3^2 z_1^2 - b_{-1} b_{-2}^2 b_{-3}^2 z_1^2)/(b_1 b_2^3 b_3^2 - b_2 b_{-1} b_{-2}^2 b_{-3}^2)$$

$$a_{1,3} = (z_2 b_2^2 b_3 b_{-1} b_{-2} + b_{-1} b_{-2}^3 b_{-3} z_3)/(b_1 b_2^3 b_3^2 - b_2 b_{-1} b_{-2}^2 b_{-3}^2)$$

$$a_{2,1} = (-z_2 b_1 b_2^2 b_3^2 + z_2 b_{-1} b_{-2}^2 b_{-3}^2 + b_1 b_3 b_{-2}^2 b_{-3} z_3)/(b_1 b_2^2 b_3^2 b_{-1} - b_{-1}^2 b_{-2}^2 b_{-3}^2)$$

$$a_{2,2} = (z_2 b_1^2 b_2 b_3^2 z_1 + z_2 b_1^2 b_3^2 b_{-2} z_3)/(b_1 b_2^2 b_3^2 b_{-1} - b_{-1}^2 b_{-2}^2 b_{-3}^2)$$

$$a_{2,3} = (z_2 b_1 b_{-1} b_{-2}^2 b_{-3} + b_1^2 b_3 b_{-2}^2 z_3)/(b_1 b_2^2 b_3^2 b_{-1} - b_{-1}^2 b_{-2}^2 b_{-3}^2)$$

$$a_{3,1} = (b_2^4 b_3^2 z_3)/(b_1 b_2^2 b_3^2 b_{-2}^2 - b_{-1} b_{-2}^4 b_{-3}^2)$$

$$a_{3,2} = (z_2 b_2^3 b_3 b_{-1} b_{-3} z_1 + z_2 b_2^2 b_3 b_{-1} b_{-2} b_{-3} z_3 - b_1 b_2^3 b_3^2 z_3 z_1 - b_1 b_2^2 b_3^2 b_{-2} z_3^2 + b_2 b_{-1} b_{-2}^2 z_3 z_1 + b_{-1} b_{-2}^2 b_{-3}^2 z_3^2)/(b_1 b_2^2 b_3^2 b_{-2}^2 - b_{-1} b_{-2}^4 b_{-3}^2)$$

$$a_{3,3} = (z_2 b_2^4 b_3 b_{-1} + b_2^2 b_{-1} b_{-2}^2 b_{-3} z_3)/(b_1 b_2^2 b_3^2 b_{-2}^2 - b_{-1} b_{-2}^4 b_{-3}^2)$$

7. Concluding remarks. It is possible to obtain the relations like kinetic polynomial for multiroute mechanism. After the elimination of unknown intermediate concentrations one can reduce the model to the system of polynomials in the reaction path rates (or the rates of reaction of certain observable species). The structure of coefficients in this case however may be rather complex than for single-route mechanism. Consider, for example, the two-route mechanism

(7.1)

$$1) \quad A_2 + 2Z \underset{k_{-1}}{\overset{k_1}{\rightleftharpoons}} 2AZ \qquad |1 \quad 1|$$

$$2) \quad B + Z \underset{k_{-2}}{\overset{k_2}{\rightleftharpoons}} BZ \qquad |2 \quad 2|$$

$$3) \quad AZ + BZ \underset{k_{-3}}{\overset{k_3}{\rightleftharpoons}} 2Z + AB \quad |2 \quad 0|$$

$$4) \quad AZ + B \underset{k_{-4}}{\overset{k_4}{\rightleftharpoons}} Z + AB \quad |0 \quad 2|$$

$$\overline{A_2 + 2B \rightleftharpoons 2AB}$$

that is the combination of adsorption and impact mechanism (3.6) and (3.1). The polynomial in the rate of formation of reaction product has fourth degree like (3.9). Its free member is

(7.2)
$$B_0 = 4b_{-2}^2(b_{-1}(b_1 b_2^2 b_3^2 - b_{-1} b_{-2}^2 b_{-3}^2) + 2b_{-1} b_1 (b_{-2} + b_3 + b_4)(b_4 b_{-2} b_{-3} - b_{-4} b_2 b_3) +$$
$$2b_{-1}(b_2 + b_{-2} - b_{-4})(b_1 b_2 b_3 b_4 - b_{-1} b_{-2} b_{-3} b_{-4}) + (b_1 b_4^2 - b_{-1} b_{-4}^2)$$
$$(b_{-1}(b_2 + b_{-2} - b_{-4})^2 - b_1(b_{-2} + b_3 + b_4)^2))$$

---

[15]computers IBM PC AT, PS2.

We should note that there are four terms in expression (7.2). The 1st and 4th correspond the basic routes in the mechanism (7.1). The 2nd and 3rd correspond their linear combinations. Thus the structure of the free term for multiroute case is more complex than in single-route one (see Theorem 2.1). Nevertheless the expression (7.2) is thermodynamically correct. It follows from conditions of equilibrium (see [14]) that $k_4/k_{-4} = k_2k_3/k_{-2}k_{-3}$. Therefore $b_4b_{-2}b_{-3} = b_{-4}b_3b_3$ for appropriate values of c_{A_2}, c_B, and c_{AB}. Thus expression (7.2) can be written as $B_0 = B_0'(k_1k_2^2k_3^2c_{A_2}c_B^2 - k_{-1}k_{-2}^2k_{-3}^2c_{AB}^2)$ similar to that of single-route mechanism (3.1) and (3.6).

Kinetic polynomial is the general form of complex reaction kinetic equation. As a rule it cannot be simplified to the explicit equation like (1.2) suggested by Boreskov and Horiuti. The last is valid only in limiting cases: linear mechanism (i), existence of rate-determining step (ii), neighbourhood of equilibrium (iii). These cases have been analysed on the basis of kinetic polynomial and there were obtained explicit analytical expressions here. Kinetic polynomial can find various applications. It simplifies the analysis of various limiting cases, facilitates the solution of the inverse kinetic problem. It seems advisable to apply it in describing complicated kinetic behaviour (e.g. multiplicity of steady states). It can also be used as a correct regression model. (The known regression models do not satisfy the "equilibrium test"). Complex expressions of the coefficients of kinetic polynomial can be effectively derived by the methods of computer algebra.

We have proved recently that forms like kinetic polynomial can be obtained not only in the case of ideal kinetic low for the elementary step (MAL), but also in the case of non-ideal kinetics or for non-isothermal conditions. There are no however the explicit polynomial expressions for their coefficients as in the case of ideal models. But these generalisations as well as the problem of variable elimination in the case of unsteady state kinetic models are out the scope of presented paper.

REFERENCES

1. J. HORIUTI, *Theory of reaction rates as based on the stoichiometric number concept*, Ann. New York Acad. Sci. 213 (1973), pp. 5–30.
2. M.I. TEMKIN, *The kinetics of some industrial heterogeneous catalytic reactions*, Adv. in Catalysis 28 (1979), pp. 173–291.
3. J. HORIUTI, "Reaction kinetics", Iwanami Book Co., Tokyo (1940).
4. G.K. BORESKOV, *The relation between molecularity and activation energies of reaction in forward and back directions*, Z. Fiz. Chim. 19 (1945), pp. 92–95.
5. S. DE GROOT AND P. MAZUR, Nonequilibrium Thermodynamics", Mir, Moscow (1964).
6. T. NACAMURA, *Note on chemical kinetics in the neighbourhood of equilibrium*, J. Res. Inst. Catal., Hokkaido Univ. 6 (1958), pp. 20–27.
7. M. BOUDART, "Kinetics of Chemical Processes", Englewood Cliffs, New Jersey (1968).
8. S.L. KIPERMAN, "Introduction to the Kinetics of Heterogeneous Catalytic Reactions", Nauka, Moscow (1964).
9. J. HAPPEL, *A rate expression in heterogeneous catalysis*, Chem. Eng. Sci. 22 (1967), pp. 479–480.
10. J. HAPPEL, *Comments on Huriuti's stoichiometric number concept*, J. Res. Inst. Catal., Hokkaido Univ. 28 (1980), pp. 185–188.
11. M. BOUDART, D.G. LOFFLER AND J.C. GOTTIFREDI, *Comments on the linear relation between reaction rate and affinity*, Int. J. Chem. Kinetics 17 (1985), pp. 1119–1123.

12. M.S. SPENCER, *Thermodynamic constrains on multicomponent catalytic systems. II. Limits to pseudo-mass-action kinetics*, J. Catalysis 94 (1985), pp. 148–154.

13. E. KING AND C. ALTMAN, *A schematic method of deriving the rate laws for enzyme-catalyzed reactions*, J. Phys. Chem. 60 (1956), pp. 1375–1381.

14. G.S. YABLONSKII, V.I. BYKOV AND A.N. GORBAN, "Kinetic Models of catalytic Reactions", Nauka, Novosibirsk (1983).

15. A.G. KUROSH, "Highter Algebra", Mir, Moscow (1972).

16. B.L. VAN DER WAERDEN, "Modern Algebra, Part 2", Ungar, New York (1970).

17. G.S. YABLONSKII, M.Z. LAZMAN AND V.I. BYKOV, *Stoichiometric number, molecularity and multiplicity*, React. Kinet. Catal. Lett. 20 (1982), pp. 73–77.

18. M.Z. LAZMAN, G.S. YABLONSKII, AND V.I. BYKOV, *Steady-state Kinetic equation. Non-linear single pathway mechanism*, Sov. J. Chem. Phys. 2 (1985), pp. 404–418.

19. M.Z. LAZMAN, G.S. YABLONSKII, AND V.I. BYKOV, *Steady-state Kinetic equation. Adsorption mechanism of a catalytic reaction*, Sov. J. Chem. Phys. 2 (1985), pp. 693–703.

20. M.Z. LAZMAN, S.I. SPIVAK AND G.S. YABLONSKII, *Kinetic polynomial and the problem of determining relations between kinetic constants when solving the inverse problem*, Sov. J. Chem. Phys. 4 (1987), pp. 781–789.

21. M.Z. LAZMAN, G.S. YABLONSKII, G.M. VINOGRADOVA AND L.N. ROMANOV, *Application of the kinetic polynomial to describe steady-state dependence of the reaction rate*, Sov. J. Chem. Phys. 4 (1987), pp. 1121–1134.

22. V.I. BYKOV, A.M. KYTMANOV, M.Z. LAZMAN, AND G.S. YABLONSKII, *Resultant of quasi-steady-state equations for a single-route n-stage mechanism*, Khim. Fiz. 6 (1987), pp. 1549–1554.

23. V.I. BYKOV, A.M. KYTMANOV, M.Z. LAZMAN, AND G.S. YABLONSKII, *Kinetic polynomial for one-route n-stage catalytic reaction*, In: Mathematical Problems of Chemical Kinetics, Nauka, Novosibirsk (1989), pp. 125–149.

24. L.A. AIZENBERG AND A.P. YUZHAKOV, "Integral representations and residues in multi-dimensional complex analysis", Nauka, Novosibirsk (1979).

25. D.N. BERNSTAIN, *The number of the roots of the system of equations*, Funct. Anal. Appl. 9 (1975), pp. 1–4.

26. V.I. ARNOL'D, A.N. VARCHENKO, AND S.M. GUSEIN-ZADE, "Singularities of differentiating mappings", Nauka, Moscow (1982).

27. M.Z. LAZMAN, "Study of nonlinear kinetic models of heterogeneous catalytic reactions (Dissertation thesis)", Institute of Catalysis, Novosibirsk (1986).

28. M.I. TEMKIN, *The kinetics of heterogeneous catalytic reactions*, Zh. D.I. Mendeleev Vses. Khim. 20 (1975), pp. 7–14.

29. G.S. YABLONSKII, V.I. BYKOV AND V.I. ELOKHIN, "Kinetics of model Reactions of Heterogeneous Catalysis", Nauka, Novosibirsk (1983).

30. T. POSTON AND I. STEWART, "Catastrophe Theory and Its Applications", Pitman, London (1978).

31. M.Z. LAZMAN, G.S. YABLONSKII, AND V.A. SOBYANIN, *Interpretation of breaks on kinetic curves*, Kinet. Catal. 27 (1986), pp. 57–63.

32. M. VLAD AND E. SEGAL, *On the kinetic model of the rate-determining step. 1*, Rev. Roumaine de Chimie 24 (1979), pp. 799–805.

33. M.M. VEINBERG AND V.A. TRENOGIN, "Theory of branching of the nonlinear equations solutions", Nauka, Moscow (1978).

34. L.A. AIZENBERG, V.I. BYKOV, A.M. KYTMANOV AND G.S. YABLONSKII, *Search for all steady-states of chemical kinetic equations with the modified method of elimination. I. Algorithm, II. Application*, Chem. Eng. Sci. 38 (1983), pp. 1555–1568.

35. G.E. FORSITHE, M. MALCOLM AND C.B. MOLER, "Computer Methods for Mathematical Computations", Englewood Cliffs, New York (1977).

36. C.L. LAWSON AND R.J. HANSON, Solving Least Squares Problems", Englewood Cliffs, New York (1974).

37. L.N. ROMANOV, *On recovery of a functional dependence by steep ordering*, Preprint No. 372, Computer Centre USSR Acad. Sci., Novosibirsk (1982).

38. G.P. MATHUR AND G. THODOS, *Initial rate approach in the kinetics of heterogeneous catalytic reactions – an experimental investigation on the sulphur dioxide oxidation reaction*, Chem. Eng. Sci. 21 (1966), pp. 1191–1200.

39. R.P.L. ABSIL, J.B. BUTT AND J.S. DRANOFF, *On the estimation of catalytic rate equation parameters*, J. Catalysis 87 (1984), pp. 530–535.

40. S.I. SPIVAK AND V.G. GORSKII, *About completeness of available kinetic data when determining kinetic constants of complex chemical reactions*, Khim, Fiz. 1 (1982), pp. 237–243.

41. V.A. EVSTIGNEEV AND G.S. YABLONSKII, *Non-identifiability of kinetic model parameters as a consequence of non-Hamiltonial structure of complex reaction graph*, Teor. i Eksper. Khimia 6 (1982), pp. 688–694.

42. V.I. BYKOV AND A.M. KYTMANOV, *An algorithm of construction of kinetic polynomial coefficients for non-linear single-route mechanism of catalytic reaction*, Preprint No 40M, Institute of Physics USSR Acad. Sci., Krasnoyrsk (1987).

43. V.I. BYKOV AND A.M. KYTMANOV, *About one modification of the method of non-linear algebraic equation system resultant construction*, Preprint No 44M, Institute of Physics USSR Acad. Sci., Krasnoyrsk (1988).

44. B. BUCHBERGER, *Grobner bases: an algorithmic method in polynomial ideal theory*, CAMP -Publ. Nr 83–290 (1983).

45. F. WINKLER, B. BUCHBERGER, F. LICHTENBERGER AND H. ROLLETSCHK, *An algorithm for constructing canonical base of polynomial ideals*, ACM Trans. Math. Software 11 (1985), pp. 66–78.

46. A.C. HEARN, "REDUCE user's manual. Version 3.2", Rand Corporation, Santa Monica (1985).

47. V.I. BYKOV, A.M. KYTMANOV AND M.Z. LAZMAN, "Methods of elimination in computer algebra of polynomials", Nauka, Novosibirsk, (to appear) (1991).

CONVERGENCE OF TRAVELLING WAVES FOR PHASE FIELD EQUATIONS TO SHARP INTERFACE MODELS IN THE SINGULAR LIMIT

YASUMASA NISHIURA* AND GUNDUZ CAGINALP†

Abstract. We show the convergence of travelling waves for the phase field equations to that of a modified Stefan model in an appropriate singular limit. Finite surface tension effect is crucial to prove this convergence.

§1. Introduction. It is known in [1], [4] that, by using a formal asymptotic analysis, the phase field equations approach a sharp interface model with curvature and dynamic cooling effects in an appropriate singular limit.

One of the most interesting questions from dynamical point of view is that how is the relation between the solutions of the phase field model and those of the limiting sharp interface model. For the steady state problems, it has proven in [3] under various conditions that the steady state phase field equations converge to the steady state modified Stefan problem.

In this note we present the first rigorous convergence results in the dynamical setting. Namely, we show the existence of one-dimensional travelling waves for the phase field equations (1.1), (1.2) which converge to that of the modified Stefan model (1.6). The detailed proof is given in [5].

The phase field equations can be written as

$$\text{(1.1)} \qquad \alpha \xi^2 \varphi_t == \xi^2 \Delta \varphi + a^{-1} g(\varphi) + 2u$$

$$\text{(1.2)} \qquad u_t + \frac{l}{2} \varphi_t = K \Delta u$$

subject to appropriate initial and boundary conditions, where l and K are dimensionless latent heat and diffusivity, respectively. The function g is a derivative of a symmetric double well potential with minima at ± 1, for instance, $g(\varphi) = \frac{1}{2}(\varphi - \varphi^3)$. The parameters α, ξ and a are related to the relaxation time, strength of microscopic interactions and the depth of the double well potential, respectively. ξ and a are usually taken to be small.

For our purpose, it is convenient to define two parameters

$$\text{(1.3)} \qquad \varepsilon \equiv \xi a^{1/2} \quad , \quad \sigma_0 = m \xi a^{-1/2} ,$$

*Mathematics Department, Hiroshima University, Hiroshima 730, Japan
†Mathematics Department, University of Pittsburgh, PA 15620 USA

where ε measures the width of the transition layer for φ, σ_0 is related to the surface tension, and m is a positive constant defined by $m \equiv \|\Phi^0_y\|^2_{L^2(\mathbf{R})}$ (see (2.13) for Φ^0). The equation (1.1) becomes

(1.4)
$$\alpha \varepsilon^2 \varphi_t = \varepsilon^2 \Delta \varphi + g(\varphi) + \frac{2\varepsilon m}{\sigma_0} u \ .$$

We are concerned with the singular limit where ξ and a tend to zero, but $\xi a^{-1/2}$ remains finite, namely

(1.5)
$$\varepsilon \downarrow 0 \quad , \quad \sigma_0 = \quad \text{fixed} \ > 0 \ .$$

That is, one can allow the interfacial thickness to vanish while the surface tension remains fixed. In this scaling limit, it was shown in [1] by a formal asymptotic analysis that the phase field equations approach the following sharp interface model (called the "modified Stefan model" in [1]):

$$u_t = K \Delta u$$

(1.6)
$$\left. \begin{array}{c} lv = K[\nabla u \cdot n]^-_+ \\[2mm] u = -\dfrac{\sigma_0}{\Delta S} \kappa - \alpha \dfrac{\sigma_0}{\Delta S} v \end{array} \right\} \quad \text{on } \Gamma \ ,$$

where Γ denotes the interface, v is the normal velocity of the interface, $[\ ]^-_+$ denotes the jump in the normal derivative of u from solid to liquid, κ is the sum of principal curvatures and ΔS is the entropy difference between the two phases. It is not difficult to verify [2], [7] that (1.6) has the following travelling wave solutions with velocity c^* and $u(t, +\infty) = u_{\text{cool}}$:

$$u(t, x) = \begin{cases} u_{\text{cool}} + le^{-\frac{c^*}{K}(x - c^* t)} & x > c^* t \\[2mm] u_{\text{cool}} + l & x \le c^* t \end{cases}$$

(1.7)
$$c^* = -\frac{4}{\alpha \sigma_0} (u_{\text{cool}} + l) \ .$$

For travelling waves to the phase field equations, we impose the boundary conditions

(1.8)
$$u(t, -\infty) = u^\varepsilon_S \ , \ \varphi(t, -\infty) = \varphi^\varepsilon_{-\infty}$$

$$u(t, \infty) = u_{\text{cool}} < 0, \ \varphi(t, \infty) = \varphi^\varepsilon_\infty \ ,$$

where $\varphi^\varepsilon_{-\infty}$ is the left most root of $g(\varphi) + \dfrac{2\varepsilon m}{\sigma_0} u^\varepsilon_S = 0$, and $\varphi^\varepsilon_\infty$ is the right most root of $g(\varphi) + \dfrac{2\varepsilon m}{\sigma_0} u_{\text{cool}} = 0$. Note that the above boundary conditions are not independently given, in fact, it turns out in §2 that once u_{cool} is given, other data are uniquely determined as functions of ε. We also assume that

(1.9)
$$u_{\text{cool}} + l < 0$$

saying that the supercooling temperature u_{cool} is sufficiently low compared with the latent heat. This ensures the positivity of the travelling velocity.

Our main result is the following.

THEOREM [5]. *In the scaling limit (1.5) and under (1.9), the phase field equations (1.2), (1.4) has an ε-family of one-dimensional travelling waves $(\varphi^\varepsilon, u^\varepsilon)$ with velocity $c = c(\varepsilon)$ satisfying (1.8). Moreover, as $\varepsilon \downarrow 0, (\varphi^\varepsilon, u^\varepsilon)$ converge to the travelling wave (1.7) of the modified Stefan model. More precisely, $c(\varepsilon)$ is a continuous function of ε up to $\varepsilon = 0$ satisfying $\lim_{\varepsilon \downarrow 0} c(\varepsilon) = c^*$, and $(\varphi^\varepsilon, u^\varepsilon)$ has the following limiting behavior for any $A > 0$:*

$$\lim_{\varepsilon \downarrow 0} \varphi^\varepsilon = \begin{cases} -1 & \text{uniformly for } z \in (-\infty, -A) \\ 1 & \text{uniformly for } z \in (A, \infty) \end{cases}$$

$$\lim_{\varepsilon \downarrow 0} u^\varepsilon = \begin{cases} u_{\text{cool}} + l & \text{for } z \le 0 \\ u_{\text{cool}} + le^{-\frac{c^*}{K}(x - c^* t)} & \text{for } z > 0 \end{cases}$$

where the limit is uniform on $\mathbf{R}$ with respect to the travelling coordinate $z = x - c(\varepsilon)t$. Here we fix the phase of the travelling waves as

$$\varphi^\varepsilon(0) = \frac{1}{2} .$$

REMARK 1.1. In different scaling regimes where (1.1), (1.2) approach the classical Stefan model or the alternative modified Stefan model (see [1]), it can be proved [5] that the phase field equations do "not" have travelling waves under the condition (1.9).

There are several related works: travelling wave solutions to the phase field equations in a different scaling regime (i.e., $a = $ constant) has been studied numerically [9], and a bifurcation from planar to curved travelling fronts has been obtained in [8] for a different scaling regime (i.e., $a = $ constant, $\alpha = O(\xi^{-1})$).

§2. Travelling wave solutions to the phase field model and its singular limit. Introducing the moving coordinate

$$(2.1) \qquad z = x - ct ,$$

equations (1.2), (1.4) become

$$(2.2) \qquad \varepsilon^2 \varphi_{zz} + \alpha \varepsilon^2 c \varphi_z + g(\varphi) + \frac{2\varepsilon m}{\sigma_0} u = 0$$

$$(2.3) \qquad K u_{zz} + c(u_z + \frac{l}{2} \varphi_z) = 0 .$$

Integrating (2.3) from $z = -\infty$ (see (1.8)), we have

$$(2.4) \qquad K u_z + c\{(u - u_S^\varepsilon) + \frac{l}{2} (\varphi - \varphi_{-\infty}^\varepsilon)\} = 0 .$$

The associated first order system (2.2), (2.4) is given by

(2.5)
$$
\begin{cases}
\varepsilon\varphi_z = \chi \\[2mm]
\varepsilon\chi_z = -\alpha\varepsilon c\chi - g(\varphi) - \dfrac{2\varepsilon m}{\sigma_0}\, u \\[4mm]
u_z = -\dfrac{c}{K}\,\{u - u_s^\varepsilon\} + \dfrac{l}{2}(\varphi - \varphi_{-\infty}^\varepsilon)\} \ .
\end{cases}
$$

The travelling wave for (2.2), (2.4) corresponds to a heteroclinic orbit of (2.5) which connects two distinct equilibria. The equilibria of (2.5) are given by the intersection of two curves

(2.6)
$$
g(\varphi) + \frac{2\varepsilon m}{\sigma_0} u = 0 \quad \text{and} \quad u - u_s^\varepsilon + \frac{l}{2}\,(\varphi - \varphi_{-\infty}^\varepsilon) = 0 \ .
$$

It is evident from (2.6) that once the supercooling temperature $u_{\mathrm{cool}}(< 0)$ is given, all the other data $\varphi_\infty^\varepsilon, u_s^\varepsilon, \varphi_{-\infty}^\varepsilon$ are uniquely determined as intersecting points of the cubic curve and the straight line with slope $l/2$.

We are interested in a heteroclinic orbit connecting $(\varphi_\infty^\varepsilon, 0, u_{\mathrm{cool}})$ to $(\varphi_{-\infty}^\varepsilon, 0, u_S^\varepsilon)$. Note that

(2.7)
$$
\varphi_{\pm\infty}^\varepsilon = \pm 1 + \varepsilon\eta_\pm(\varepsilon) \ , \quad u_S^\varepsilon = u_{\mathrm{cool}} + l + \varepsilon\rho(\varepsilon) \ ,
$$

hold where $\eta_\pm(\varepsilon)$ and $\rho(\varepsilon)$ are smooth and bounded up to $\varepsilon = 0$.

The strategy is as follows: first we solve (2.4) with respect to u, and substitute the resulting formula into (2.2) to obtain a single equation for φ. Applying the alternative method to it, the whole problem is reduced to solving a bifurcation equation with respect to the velocity c and ε. The key is that the temperature field u does "not" have a limit in $\mathcal{B}(\mathbf{R})$-space as $\varepsilon \downarrow 0$, however the scaled bifurcation equation remains valid up to $\varepsilon = 0$.

For our purpose, it is convenient to use the stretched coordinate $y(\equiv z/\varepsilon)$ and the "inner variables":

(2.8)
$$
\Phi(y) = \varphi(\varepsilon y) \ , \quad U(y) \equiv u(\varepsilon y) \ .
$$

Equations (2.2), (2.4) become

(2.9)
$$
\Phi_{yy} + \alpha\varepsilon c\Phi_y + g(\Phi) + \frac{2\varepsilon m}{\sigma_0} U = 0
$$

(2.10)
$$
KU_y + c\{(U - u_s^\varepsilon) + \frac{l}{2}(\Phi - \varphi_{-\infty}^\varepsilon)\} = 0 \ .
$$

Solving (2.10) under (1.8) and substituting it into (2.9), we have

(2.11)

$$\Phi_{yy} + \alpha \varepsilon c \Phi_y + g(\Phi)$$
$$+ \frac{2\varepsilon m}{\sigma_0} \left\{ u_s^\varepsilon - \frac{\varepsilon c}{K} \int_{-\infty}^{y} e^{-\frac{\varepsilon c}{K}(y-s)} \frac{l}{2} \left(\Phi(s) - \varphi_{-\infty}^\varepsilon \right) ds \right\} = 0$$

(2.12)
$$\Phi(+\infty) = \varphi_{+\infty}^\varepsilon , \quad \Phi_y(\pm\infty) = 0 .$$

Letting $\varepsilon \downarrow 0$ and using (2.7), we have the reduced problem

(2.13)
$$\Phi_{yy} + g(\Phi) = 0 , \quad \Phi(\pm\infty) = \pm 1$$

which has a unique (up to translation) strictly increasing solution $\Phi^0(y)$ decaying exponentially at infinity. It is natural to seek a solution of the form

(2.14)
$$\Phi = \Phi^0 + \Psi .$$

The equation for Φ becomes

(2.15$_a$)

$$\Psi_{yy} + g'(\Phi^0)\Psi + \alpha \varepsilon c \Phi_y^0 + \alpha \varepsilon c \Psi_y + G(\Psi, \Phi^0)$$
$$+ \frac{2\varepsilon m}{\sigma_0} \left\{ u_s^\varepsilon - \frac{\varepsilon c}{K} \int_{-\infty}^{y} e^{-\frac{\varepsilon c}{K}(y-s)} \frac{l}{2} \left(\Phi^0(s) + \Psi - \varphi_{-\infty}^\varepsilon \right) ds \right\} = 0 ,$$

(2.15$_b$)
$$\Psi(\pm\infty) = \varepsilon \eta_\pm(\varepsilon) , \quad \Psi_y(\pm\infty) = 0 ,$$

where $G(\Psi, \Phi^0) \equiv g(\Phi^0 + \Psi) - g(\Phi^0) - g'(\Phi^0)\Psi$. We seek a solution Ψ of (2.15) in the $\mathcal{B}$-space, i.e.,

$$\mathcal{B}^k(\mathbb{R}) \equiv \{ f : \mathbb{R} \to \mathbb{R} \quad \text{has bounded and}$$

$$\text{continuous derivatives up to the k-th order} \}$$

with the norm $\|f\|_k = \sum_{j=0}^{k} \sup_{z \in \mathbb{R}} |D^j f(z)|$.

REMARK 2.1. The reason why we do "not" put $\Phi^0 + \varepsilon\Psi$ in (2.14) is that if we do so, the resulting Ψ does "not" have a limit in $\mathcal{B}^0$-space as $\varepsilon \to 0$.

The principal linear part of (2.15)$_a$ is $L \equiv \dfrac{d^2}{dy^2} + g'(\Phi^0)$ which has one-dimensional kernel in $\mathcal{B}^0(\mathbb{R})$ spanned by $\{\Phi_y^0\}$. This is a typical situation where we can use the alternative method (see, for instance, [6, Chap. 2]). Applying the usual procedures of the alternative method to (2.15), we see that the problem is reduced to solving the "bifurcation equation" with respect to c and ε:

(2.16)

$$B(c,\varepsilon) \equiv \Big\langle \alpha\varepsilon c\Phi_y^0 + \alpha\varepsilon c\Psi_y(c,\varepsilon) + G(\Psi(c,\varepsilon),\Phi^0)$$

$$+ \frac{2\varepsilon m}{\sigma_0} \left\{ u_s^\varepsilon - \frac{\varepsilon c}{K} \int_{-\infty}^{y} e^{-\frac{\varepsilon c}{K}(y-s)} \frac{l}{2} (\Phi^0(s) + \Psi(c,\varepsilon) - \varphi_{-\infty}^\varepsilon) ds \right\} , \Phi_y^0 \Big\rangle = 0 ,$$

where $\Psi(c,\varepsilon) \in \mathcal{N}(L)^\perp$ is the unique solution of $(2.15)_a$ projected to the range space of L. Unfortunately (2.16) is degenerate in the sense that $B(c,0) = 0 = \frac{\partial B}{\partial c}(c,0)$, because we put $\Phi = \Phi^0 + \Psi$ (not $\Phi^0 + \varepsilon\Psi$) as in (2.14) which is necessary to obtain a solution Ψ in $\mathcal{B}^0$-space. However it can be proved that the scaled bifurcation equation

(2.17)
$$\widetilde{B}(c,\varepsilon) \equiv B(c,\varepsilon)/\varepsilon$$

is well-defined up to $\varepsilon = 0$ owing to the exponentially decaying property of Φ_y^0. Namely, although the integral

(2.18)
$$\frac{\varepsilon c}{K} \int_{-\infty}^{y} e^{-\frac{\varepsilon c}{K}(y-s)} \frac{l}{2} (\Phi^0(s) + \Psi(c,\varepsilon) - \varphi_{-\infty}^\varepsilon) ds$$

does not have a limit in $\mathcal{B}^0(\mathbf{R})$-space as $\varepsilon \downarrow 0$, the inner product with Φ_y^0 has a definite limit $(= 0)$, since (2.18) converges to zero uniformly on any "compact" set of $\mathbf{R}$ when $\varepsilon \downarrow 0$. In fact, we can prove

PROPOSITION 2.1. *There is a positive ε_0 such that $\widetilde{B}(c,\varepsilon)$ and $(\partial\widetilde{B}/\partial c)(c,\varepsilon)$ are well defined and continuous for $(c,\varepsilon) \in I \times [0,\varepsilon_0]$, where I is an arbitrary given bounded interval in $\mathbf{R}^+$. Moreover, one has*

$$\widetilde{B}(c,0) = \alpha cm + \frac{4m}{\sigma_0}(u_{\text{cool}} + l)$$

$$\frac{\partial\widetilde{B}}{\partial c}(c,0) = \alpha m > 0 ,$$

where $m = \|\Phi_y^0\|_{L^2(\mathbf{R})}^2$.

Hence it follows from Proposition 2.1 and the implicit function theorem that there is a unique continuous function $c = c(\varepsilon)$ with limiting velocity

$$c^* \equiv c(0) = -\frac{4}{\alpha\sigma_0}(u_{\text{cool}} + l) > 0$$

satisfying $\widetilde{B}(c(\varepsilon),\varepsilon) = 0$ for $\varepsilon \in [0,\varepsilon_0]$. This leads to Theorem in §1.

REFERENCES

[1] G. CAGINALP, *Stefan and Hele–Shaw type models as asymptotic limits of the phase field equations*, Physical Review A 39 (1989), 5887–5896.

[2] G. CAGINALP AND J. CHADAM, *Stability of interfaces with velocity correction term*, to appear in Rocky Mountain J. of Math.

[3] G. CAGINALP AND P.C. FIFE, *Elliptic problems involving phase boundaries satisfying a curvature condition*, IMA J. of Applied Math. 38 (1987), 195–217.

[4] G. CAGINALP AND P.C. FIFE, *Dynamics of layered interfaces arising from phase boundaries*, SIAM J. Appl. Math 48 (1988), 506–518.

[5] G. CAGINALP AND Y. NISHIURA, *The existence of travelling waves for phase field equations and convergence to sharp interface models in the singular limit*, To appear in Quarterly of Appl. Math.

[6] S–N. CHOW AND J. HALE, *Methods of Bifurcation Theory*, Springer, Berlin (1982).

[7] J.N. DEWYNNE, S.D. HOWISON, J.R. OCKENDON, W. XIE, *Asymptotic behavior of solutions to the Stefan problem with a kinetic condition at the free boundary*, J. Austral. Math. Soc. Ser. B 31 (1989), 81–96.

[8] H. FUJII, Y. NISHIURA, M. MIMURA, R. KOBAYASHI, *Existence of curved fronts for the phase field model*, In preparation.

[9] J.W. WILDER, *Travelling wave solutions for interfaces arising from phase boundaries based on a phase field model*, Rensselaer Polytechnics Inst. Preprint.

STANDING AND PROPAGATING TEMPERATURE WAVES ON ELECTRICALLY HEATED CATALYTIC SURFACES*

GEORGIOS PHILIPPOU AND DAN LUSS†

Abstract. Reaction rates are often measured using a catalytic wire or ribbon, the resistance (average temperature) of which is kept at a preset value via electrical heating. Previous investigators assumed that the wire temperature was uniform. An IR-thermal imager shows that some temperature profiles have the shape of a stationary standing wave. A bifurcation map describes the organization of the regions with these standing wave temperature profiles. In some cases, the high temperature wave moves back and forth on the ribbon, with continuous changes in its shape. This leads to both oscillatory and chaotic changes in the overall rate of heat generated by the reaction. The dynamic behavior of the overall reaction rate is different and less regular than that of local temperatures on the ribbon. The power spectrum of the overall rate of heat generation decays exponentially, while that of the local temperatures decays as a power law. Ignoring the nonuniform nature of the temperature of the ribbon may lead to severe pitfalls in the determination of the kinetic rate expression and/or its parameters

Key words. catalytic surfaces, stationary temperature waves, propagating temperature waves.

1. Introduction. A large number of spatio-temporal patterns has been observed in chemically reacting liquid phase systems, especially with the Belousov-Zhabotinskii and the glycolysis reactions[1]. These include uni-dimensional excitable and train waves, planar target patterns and spiral waves, and three dimensional movements of scroll-waves. In recent years similar spatio-temporal patterns were found in some heterogeneous catalytic systems. The group of Ertl has been able to observe changes in surface structures or surface coverage on single crystals using scanning low energy electron-diffraction, scanning photoemission and photoemission electron microscopy. They observed propagating and standing waves, rotating spirals and chaos during the low pressure oxidation of carbon monoxide[2,3]. The groups of Schmitz[4], Wolf[5] and Schmidt[6] observed temperature waves on heterogeneous catalytic surfaces on which chemical reactions were carried out at atmospheric conditions. We explain here how electrical heating of catalytic surfaces may generate both stationary and propagating temperature waves.

2. Standing Temperature Waves on Catalytic Surfaces. Electrically heating wires and ribbons have been used for many years to study the multiplicity features of catalytic systems. The electrical heating is used to keep the resistance and hence the average temperature of the wire at a preset value. It was assumed in most previous studies that the temperature of these heated wires was uniform. Hence, bifurcation diagrams of the heat generated by the reaction vs. wire temperature were used to estimate the kinetic rate constants and/or the functional form of the rate expression.

Figure (1) describes a typical experimental bifurcation diagram of heat generated vs. average wire temperature obtained during the oxidation of a mixture

*We are thankful to the National Science Foundation and the Welch Foundation for support of this research.

†Department of Chemical Engineering, University of Houston, Houston, Texas, 77204-4792

containing 1%, ammonia in air. The catalyst was a 14.7 cm long, 0.05 cm wide by 0.0025 cm thick pure platinum ribbon. A hot wire anemometer (TSI, IFA-100) maintained the ribbon at a preset total resistance (and therefore constant average temperature, $\overline{T}$). The temperature profiles were measured by an infra-red imager (AGEMA thermovision 780), mounted on a motorized table and driven parallel to the ribbon. Additional details of the experimental system can be found in[7].

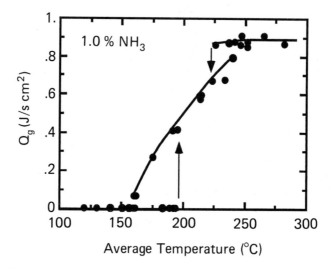

Figure 1: Bifurcation diagram of the heat generated by oxidation of a mixture containing 1% ammonia vs. average ribbon temperature.

The experiments showed that a uniform, extinguished state with a negligible rate of heat generation existed for all temperatures below 195°C. A fully ignited branch of solutions existed for all $\overline{T} > 226°C$. The temperature profiles of these states (see Fig. 2, $\overline{T} = 252$ and 282°C) were uniform with the exception of a narrow end region next to the supports.

A branch of nonuniform temperature (NUTR) states existed for all $\overline{T}$ in (160, 240°C). The temperature profiles of these states were usually non-symmetric about the center of the ribbon and consisted of two segments of different but rather uniform temperatures, separated by a narrow front. Increasing the average ribbon temperature increased the fraction maintained at the high temperature. Figure 2 describes some profiles obtained by a slow continuous increase in the average wire temperature. The standing temperature waves remained stationary for very long periods of time as any movement of the front changed the electrical resistance of the ribbon. The anemometer immediately changed the electrical current to counteract this movement and restored the temperature front to its original position. Thus,

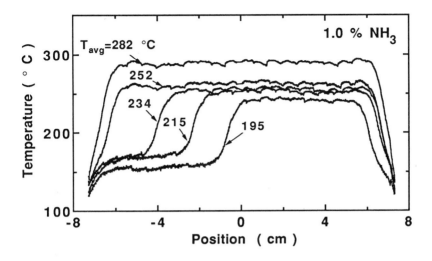

Figure 2: Temperature profiles of the ribbon obtained as its average temperature is slowly increased.

the electrical heating generated and stabilized these standing temperature waves.

The experiments showed that either an extinguished or a NUTR existed for $\overline{T}$ in (160, 195°C). The two states which can exist for 195°C are shown in Fig. 3. Similarly, either an ignited or a NUTR existed for $\overline{T}$ in (226, 240°C). Only a NUTR existed for $\overline{T}$ in (195, 226°C).

Repeated experiments showed that the branch of NUTR did not emanate from either one of the two uniform temperature branches. A sub-critical bifurcation occurred at both the ignition and extinction points of the uniform states. The exact transition from the NUTR to a branch of uniform states was difficult to locate due to the proximity of the two branches.

A bifurcation map of regions with qualitatively different states (Fig. 4) was constructed from the bifurcations diagrams for several feed concentrations. The map shows that five qualitatively different regions exist, in three of which a single state exists (either ignited, extinguished or nonuniform) while in two regions two different types of states can be obtained. We were not able to determine how the boundaries of the various regions coalesce at low ammonia concentrations, due to experimental inaccuracies in that range of operation.

In modeling this system, we assume that the local surface concentration is in equilibrium with that of the gas. A species balance gives:

$$(1.1) \qquad k_c(C_b - C_s) = r(C_s, T)$$

where k_c is the mass transfer coefficient, C_b the bulk reactant concentration, and the reaction rate, r, depends on the local concentration (C_s) and temperature (T).

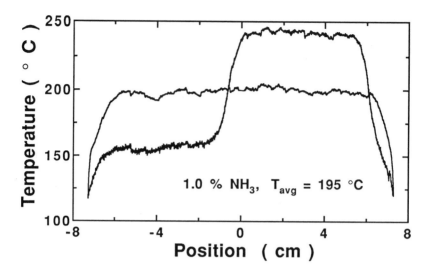

Figure 3: A uniform and NUTR temperature profiles having the same average temperature of 195°C.

This equation may be brought to the form:

$$(1.2) \qquad C_s = g(T, \mathbf{p})$$

where $\mathbf{p}$ is a vector of parameters. Thus, the local reaction rate can be expressed as a function of the local temperature. The energy balance for the ribbon is:

$$(1.3) \qquad \rho c_p A \frac{\partial T}{\partial t} = \lambda A \frac{\partial^2 T}{\partial x^2} + f(T)$$

where

$$(1.4) \qquad f(T) = a[(-\Delta H)r(T) - h(T - T_a)] + I^2 R(T)$$

and A is the cross section of the ribbon, a the perimeter of the ribbon, λ the effective thermal conductivity, h the heat transfer coefficient, I the electrical current, and R the ribbon resistance per unit length. Usually, we can express R as a linear function of the temperature, i.e.,

$$(1.5) \qquad R(T) = R(T_a)[1 + \alpha(T - T_a)]$$

It is well known that a standing temperature wave may exist in an infinite medium (one with no end effects) only if[8]

$$(1.6) \qquad \int_{T_1}^{T_2} f(T)dT = 0$$

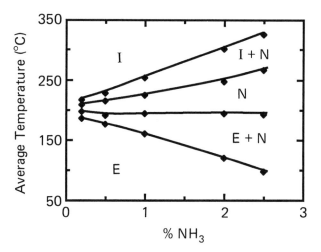

Figure 4: A bifurcation map describing regions with qualitatively different steady states; E refers to low temperature (extinguished) state, I high temperature (ignited) state, and N to a nonuniform temperature state.

where T_1 and T_2 are the uniform temperatures on both sides of the front. The uniform temperature states satisfy the relation $f(T) = 0$, which may be written as[9];

$$(1.7) \qquad\qquad Q_R = Q_g$$

where:

$$(1.8) \qquad Q_g = (-\Delta H)r(T)$$
$$Q_R = -\frac{h}{\alpha} + \left(h - \frac{\alpha R(T_a)I^2}{a}\right)\left(T - T_a + \frac{1}{\alpha}\right)$$

Q_g, the heat generated by the reaction is usually a sigmoidal function of the ribbon temperature. Q_R, the heat removal term depends also on the electrical heating and is represented by a family of straight lines originating at $T = T_a - 1/\alpha$ and $Q_R = -h/\alpha$. The intersections between Q_R and Q_g are the uniform temperature states (see Fig. 5).

Inspection of Fig. 5 shows that a stable uniform state cannot exist for any temperature in the range (L, H). An attempt to get such a state leads to formation of a NUTR with the temperatures at the two sections of the ribbon being A and C. The corresponding heat removal is the line ABC, for which the two-hatched areas ALBA and CHBC are equal, so that Eqn (1.6) is satisfied. The fraction of

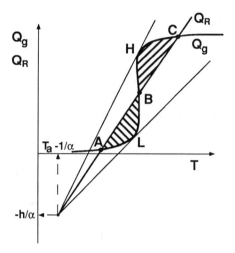

Figure 5: Graphical solution of Eqn. (1.7).

the ribbon which is at the high temperature is determined by the present average temperature.

The above analysis suggests that the current required to maintain a NUTR is independent of the average wire temperature. Experiments described in Fig. 6 indicate that the required current increases with average temperature. One possible explanation is that heat losses from the supports increase with increasing average temperature. The data show that the current needed to maintain a NUTR differs from that needed to keep a uniform state at the same average temperature.

Failure to account for the presence of NUTR may lead to severe pitfalls in the analysis of kinetic experiments. Many previous investigators used the multiplicity features of electrically heated catalytic wires to determine kinetic parameters, assuming the wires have a uniform temperature. The experiments suggest that this is a very dangerous procedure. The bifurcation map (Fig. 4) shows that if one is not accounting for the existence of the NUTR, he may interpret the data as indicating the existence of two separate regions, each with two uniform temperature stable steady states. Clearly, this is not the case. Moreover, when the boundaries of the region with the extinguished and NUTR states have different slopes in the (T, C) plane, one may erroneously conclude that a pitchfork singularity exists. We conjecture that the analysis of the multiplicity features of many electrically heated wires is erroneous as the possible existence of a NUTR was not considered.

3. Propagating Temperature Waves on Catalytic Surfaces. Experiments indicate that for certain chemical reactions and concentrations stationary temperature fronts are not formed on electrically heated catalytic surfaces. Instead, a

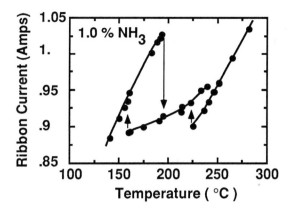

Figure 6: Electrical current needed to keep the ribbon at various average temperatures.

back and forth propagating temperature wave is obtained. Fig. 7 is a bifurcation diagram of heat generated vs. average wire temperature for the oxidation of a mixture containing 0.2% propylene in air. In this case, the overall rate of heat generation oscillated for temperature exceeding 290°C. The amplitude of these oscillations decreased with increasing temperature and became negligible at about 400°C. The heat generation is chaotic for average ribbon temperatures between 228°C and 260°C. An intermediate behavior was observed for average temperatures between 260°C and 290°C. Typical time traces of the heat generation in these three temperatures ranges are shown in Fig. 8.

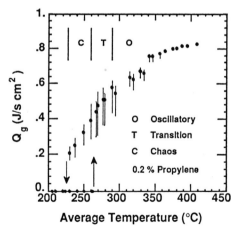

Figure 7: A bifurcation diagram of the heat generated by the oxidation of a mixture containing 0.2% propylene vs. average ribbon temperature. Bars denote amplitude of oscillations.

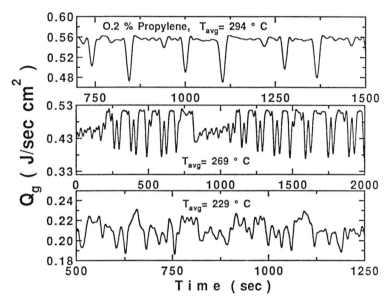

Figure 8: Transient rate of heat generated on a ribbon at the oscillatory, transition and chaotic regions for a mixture containing 0.2% propylene

Temperature profile measurements showed that the variation in the rate of heat generation were due to a back and forth movement of a high temperature zone and the temporal changes in its length (Fig. 9), as the rate of heat generation in the low temperature region is negligible. The movement of the high temperature zone in the oscillatory state consisted of five stages. Initially, the ignited section was stationary on the left side of the ribbon (profile a, Fig 9I). This zone then moved to the right, with its back moving somewhat faster than the front, reducing the length of the ignited zone and the overall rate of heat generation. The front of the zone stopped its rightwards movement before reaching the support. The wave then reversed its direction of movement, and its front moved to the left faster than its back (profile b, Fig. 9I). Eventually, the back of the wave moved faster than its front, reducing the length of the reaction zone and the overall rate. The front stopped moving leftwards before reaching the support, and the wave then started moving to the right (profile c, Fig. 9I) until it became profile a. The velocity of the moving fronts was of the order of 0.1 cm/sec.

The transient burst of irregular heat generation in the transition region (Fig. 8, $\overline{T} = 269°C$) was caused by a sudden splitting of the ignited region into two sections, which moved in the same direction at different velocities (profile d, in Fig 9II). Eventually, the two zones collided and coalesced into one (profile a, Fig 9II), returning the ribbon to an oscillatory mode.

At low wire temperatures (228°–260°C) the width of the ignited zone was smaller than that of the low temperature zone, and its movement (Fig 9III) led to a chaotic variation in the rate of the heat generation. The distinction between chaotic and oscillatory behavior is due to changes in the nature of the movement of

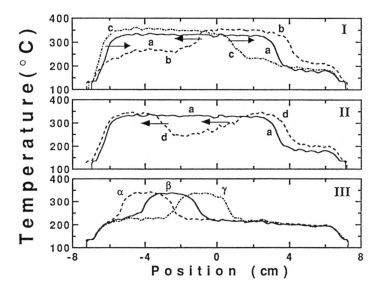

Figure 9: Transient temperature profiles observed during the oscillatory (I), transition (II) and chaotic (III) operation for a mixture containing 0.2% propylene.

the high temperature zone.

Experiments indicate that the chaotic nature of the behavior tends to become less intricate (lower correlation dimension) with increasing concentration of the propylene. At sufficiently high reactant concentrations (such as 0.5%) propylene the temperature waves become stationary.

The nature of the overall heat generation was analyzed by constructing the attractor of the dynamic behavior using the time delay method. Inspection of the attractor for 0.3% propylene (Fig. 10) shows that the chaotic motion is either around one of two unstable states, or around both states.

The chaotic temperature waves were observed at various 1 cm sections of the ribbon by keeping the IR camera stationary. Fig. 11 shows temperature profiles observed at a section located between 3.3 and 4.3 cm from the right hand support, close to the position at which the temperature wave reversed its direction. The data shows that different waves reversed the direction of movement at different positions.

The power spectrum of the chaotic heat generation signal decayed exponentially for high frequencies. Sigeti and Horsthemke[10] suggested that this implies that the system is deterministic. On the other hand, the power spectra of local temperatures decayed as a power law for high frequencies (straight line on logarithmic scales) implying a stochastic behavior. Moreover, the oscillation frequency of the

0.3 % Propylene, T$_{avg}$=225 °C

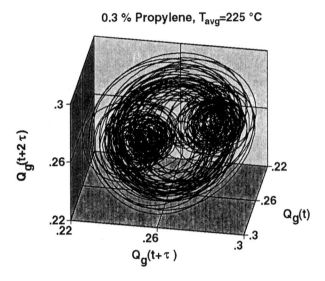

Figure 10: A reconstruction of the chaotic attractor using a time delay of 5 seconds.

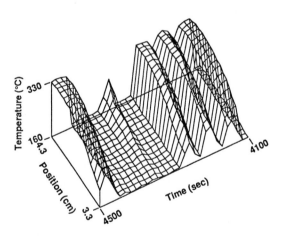

Figure 11: Temperature waves observed at a 1 cm section of the ribbon for a mixture containing 0.3% propylene and $\overline{T}$ of 243°C.

chaotic overall rate was higher than that of local temperatures. The motion of the temperature waves has many similarities to the electrical current waves observed by Lev et.al[11] during the electrochemical dissolution of nickel in a galvanostatic mode.

Many previous theoretical studies of oscillatory behavior of catalytic surfaces assumed that the surface is homogeneous. The experimental results indicate that this is often not the case, and that intricate wave motion is the cause of the complex dynamics of the overall reaction rate. It would be very useful to develop some criteria or diagnostics predicting when the observed rate is due to temporal behavior and when it is caused by a spatio-temporal process.

Propagating waves may be formed by several different mechanisms such as an instability of the high temperature state or periodic reaction induced changes in the catalytic properties of the surface. Sheintuch[12] presented a simple model in which a slow activation-deactivation process can generate spatio-temporal behavior on an electrically heated wire. It is of much interest to determine the actual chemical mechanism which creates the various dynamic features, the types of possible bifurcations and singular points, and the different qualitative behaviors. From a practical point of view, one needs to know how robust or sensitive these patterns are to slight variations in the operation conditions or catalyst. A key question is whether, the dynamic features of the ribbon may be exploited to carry out reactions, which do not proceed under stationary conditions.

REFERENCES

[1] R.J. FIELD AND M. BURGER, *Oscillations and traveling waves in chemical systems*, Wiley, New York, 1985.

[2] S. LADAS, R. IMBIHL AND G. ERTL, *Kinetic oscillations and faceting during the catalytic CO oxidation on Pt (110)*, Sur. Sci., 198 (1988), pp. 42–68.

[3] S. JAKUBITH, H.H. ROTERMUND, W. ENGEL, A. VON OERTZEN AND G. ERTL, *Spatiotemporal concentration patterns in a surface reaction: Propagating and standing waves, rotating spirals, and turbulence*, Phys. Rev. Let, 65 (1990), pp. 3013–3016.

[4] P.C. PAWLICKI AND R.A. SCHMITZ, *Spatial effects on supported catalysts*, Chem. Eng. Prog., 83(2) (1987), pp. 40–46.

[5] J.S. KELLOW AND E.E. WOLF, *Infrared thermography and effects on surface reaction dynamics during CO and ethylene oxidation on Rh/SiO_2 catalysts*, Chem. Eng. Sci., 45 (1990), pp. 2597–2602.

[6] G.A. CORDONIER AND L.D. SCHMIDT, *Thermal waves in NH_3 oxidation on a Pt wire*, Chem. Eng. Sci., 44 (1989), pp. 1983–1993.

[7] L. LOBBAN, G. PHILIPPOU AND D. LUSS, *Standing temperature waves on electrically heated catalytic ribbons*, J. Phys. Chem., 93 (1989), pp. 733–736.

[8] V. BARELKO, I.I. KUROCHKA, A.G. MERZHANOV AND K.G. SHKADINSKI, *Investigation of travelling waves on catalytic wires*, Chem. Eng. Sci., 33 (1978), pp. 805–811.

[9] M. SHEINTUCH AND J. SCHMIDT, *Observable multiplicity features of inhomogeneous solutions measured by the thermochemic method: Theory and experiments*, Chem. Eng. Comm., 44 (1986), pp. 33–52.

[10] D. SIGETI AND W. HORSTHEMKE, *High frequency power spectra for systems subject to noise*, Phys. Rev. A, 35 (1987), pp. 2276–2282.

[11] O. LEV, A. WOLFBERG, M. SHEINTUCH AND L. PISMEN, *Bifurcations to periodic and chaotic motions in anodic nickel dissolutions*, Chem. Eng. Sci., 43 (1988), pp. 1339-1353.

[12] M. SHEINTUCH, *Spatio-temporal structural of controlled catalytic wires*, Chem. Eng. Sci., 44 (1989), pp. 1081–1089.

MIXED-MODE OSCILLATIONS IN THE
NONISOTHERMAL AUTOCATALATOR

S.K. SCOTT* AND A.S. TOMLIN*

Abstract. The internal coupling of chemical and thermal feedback processes is investigated through a non-isothermal autocatalator model. The isothermal oscillatory scheme is augmented to include an exothermic "termination" step and a temperature-dependent initiation: $P \to A$ (rate $= k_0(T)p$); $A + 2B \to 3B$ (rate $= k_1 ab^2$); $A \to B$ (rate $= k_3 a$); $B \to C+$ heat (rate $= k_2 b$). The pool chemical approximation is applied to reactant P. Multiple stationary states exist for all parameter values and complex dynamic behaviour is also found over wide ranges of these parameters. Path following techniques are used to follow the periodic solutions and to find bifurcations such as period doubling cascades leading to chaos. Other types of complex dynamics such as mixed mode oscillations are categorised, although we find no quasiperiodic behaviour in the model. This gives rise to the belief that the mixed mode oscillations found, do not stem from phase locking on a torus in this instance.

1. Introduction. Mixed mode oscillations are characteristic of many nonlinear chemical systems, most notably in the Belousov–Zhabotinskii reaction [2–4,8,13–18,22,23]. Typically, mixed-mode oscillations are complex periodic waveforms comprising a certain number of large and small amplitude excurions in some particular order. As some bifurcation parameter, such as the flow rate, is varied so the waveform may change to one with fewer or more small or large peaks. The bifurcation sequences often give rise to concatenations obeying Farey arithmetic and a plot of a suitably defined "firing number" against the bifurcation parameter may lead to a Devil's staircase [17,18]. In the Belousov–Zhabotinskii reaction, these responses have been interpreted as motion on a torus [7] in the appropriate concentration phase-space.

Recently, Albahadily et al. [1] have reported mixed mode oscillations with Farey sequences etc. but which apparently do not have an underlying torus structure. Here we report similar observations for a simple (but unrelated) three-variable model, based on the cubic autocatalator scheme.

2. The model. The cubic autocatalator model [19] in its simplest, robust form involves the conversion of a pool chemical reactant P to a final product C via two intermediates A and B. The interconversion between A and B proceeds via two, simultaneous routes: an uncatalyzed step, and a process involving autoacatalysis. The model scheme can thus be represented as

$$
\begin{array}{lll}
(0) & P \to A & \text{rate} = k_0 p \\
(1) & A + 2B \to 3B & \text{rate} = k_1 ab^2 \\
(3) & A \to B & \text{rate} = k_3 a \\
(2) & B \to C & \text{rate} = k_2 b
\end{array}
$$

*Department of Physical Chemistry, University of Leeds, Leeds LS2 9JT, U.K.

(the numbering is chosen for consistency with earlier work). Many, if not most, spontaneous chemical reactions are exothermic. The heat released may cause an increase in the temperature (self-heating) of the reacting mixture. This offers another mechanism for feedback, as reaction rate constants generally vary with the local temperature. Rather than going for generality, however, we seek here to introduce self-heating and its effects in a simple form: the reactions with greatest temperature dependences (the highest activation energies) are typically the initiation steps, such as (0) in the above scheme. Similarly, the most exothermic processes are usually the chain termination steps in which intermediates are converted to stable products, such as reaction (2). For our canonical model, therefore, we assume that the rate constant k_0 has an Arrhenius temperature dependence

$$(2.1) \qquad\qquad k_0(T) = A_0 e^{-E/RT}$$

and that reaction (2) has a positive molar exothermicity, $Q = -\Delta H > 0$.

The governing equations for this scheme are: for concentrations

$$(2.2) \qquad\qquad dp/dt = -k_0(T)p$$
$$(2.3) \qquad\qquad da/dt = -k_0(T)p - k_1 ab^2 - k_3 a$$
$$(2.4) \qquad\qquad db/dt = -k_1 ab^2 + k_3 a - k_2 b$$

and for temperature

$$(2.5) \qquad\qquad c_p c_0 V(dT/dt) = VQk_2 b - \chi S(T - T_a)$$

where T_a is the temperature of the surroundings. All other terms are defined in the List of symbols at the end of this paper. The initial conditions for a given experiment might be $p = p_0, a_0 = b_0 = 0$ and $T = T_a$ at $t = 0$.

Equations (2.2–5) may be case in an equivalent dimensionless form

$$(2.6) \qquad\qquad d\mu/d\tau = -\varepsilon\mu e^{\theta}$$
$$(2.7) \qquad\qquad d\alpha/d\tau = \mu e^{\theta} - \alpha\beta^2 - \kappa_\mu \alpha$$
$$(2.8) \qquad\qquad d\beta/d\tau = \alpha\beta^2 + \kappa_\mu \alpha - \beta$$
$$(2.9) \qquad\qquad d\theta/d\tau = \delta\beta - \gamma\theta$$

Here, $\varepsilon = k_0/k_2$ will be supposed to be small compared with unity. The term e^{θ} in equation (2.7) is an approximate, but conveniently simple, representation of the Arrhenius dependence. To reduce this set to a system of just three coupled equations, we make the traditional pool chemical approximation regarding a slowly decaying, large concentration of the reactant P. Formally, we require the reduced reactant concentration μ to remain of order unity as ε tends to zero. Then the integrated form of equation (2.6), $\mu = \mu_0 e^{-\varepsilon \int e^{\theta} d\tau}$ where μ_0 is the dimensionless initial concentration of P, is approximated by $\mu = \mu_0$ for sufficiently small dimensionless time $\tau \sim 0(\varepsilon^{-1})$.

Thus, we study here the set of autonomous equations

(2.10) $$d\alpha/d\tau = \mu_0 e^\theta - \alpha\beta^2 - \kappa_u \alpha$$

(2.11) $$d\beta/d\tau = \alpha\beta^2 + \kappa_\mu \alpha - \beta$$

(2.12) $$d\theta/d\tau = \delta\beta - \gamma\theta$$

3. Results. Equations (2.10–12) involve four parameters: μ_0, δ, κ_u and γ. Of these, the dimensionless reactant concentration can be thought of as the most easily varied during a given experiment (the principal bifurcation parameter): indeed, if we remember the slow exponential decrease with finite ε, an experiment would automatically provide this variation, but we will not emphasize that point unduly here. There remain, then, three unfolding parameters, perhaps to be adjusted between successive experiments. Although, in principle, modern path following techniques [9] are able to lighten the burden of searching through such parameter spaces, the actual application is generally complicated by various operational difficulties - not least by the wide range of bifurcation phenomena that even such a simple system can support [12]. Some aspects, such as the stationary-states and the conditions for Hopf bifurcation, can be treated generally, but otherwise we choose here to report on a limited number of traverses through the behaviour of this scheme.

3.1 Stationary-states. The stationary-state solutions satisfying $d\alpha/d\tau = d\beta/d\tau = d\theta/d\tau = 0$ are given by

(3.1) $$\theta_{ss} = \delta\beta_{ss}/\gamma : \alpha_{ss} = \beta_{ss}/(\kappa_u + \beta_{ss}^2)$$

where β_{ss} is a solution of the equation

(3.2) $$\mu_0 = \beta_{ss} e^{-\delta\beta_{ss}/\gamma}.$$

The qualitative form of the stationary-state locus is the same for all μ_0, δ and γ, and is shown in figure 1. For $\mu_0 < \gamma/\delta e$ there are two stationary-state solutions, one with $\beta_{ss} < \gamma/d$ and the other with $\beta_{ss} > \gamma/d$. These two solutions merge when $\mu_0 = \gamma/\delta e$ and the system shows a thermal explosion for higher initial concentrations of the reactant.

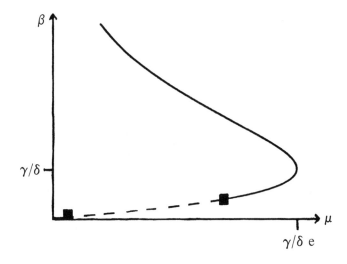

FIGURE 1. Steady state profile of β against μ, the bifurcation parameter. Marked are the Hopf bifurcation points and the dotted line between them denotes a region of dynamic instability of the stationary state.

The classic behaviour of the isothermal autocatalator is regained in either limit $\delta \to 0$ or $\gamma \to \infty$: with then the linear relationship $\beta_{ss} = \mu_0$ showing no maximum or runaway.

3.2 Hopf bifurcation. The local stability of a given stationary-state β_{ss} is determined by the sign and character of the three eigenvalues λ_{1-3} given as roots of the following cubic equation

(3.3) $$\lambda^3 + b\lambda^2 + c\lambda + d = 0$$

where

$$b = 1 + \beta_{ss}^2 + \kappa_u + \gamma - 2\beta_{ss}^2(\beta_{ss}^2 + \kappa_u)^{-1}$$
$$c = (\beta_{ss}^2 + \kappa_u)(1 + \gamma) + \gamma[1 - 2\beta_{ss}^2(\beta_{ss}^2 + \kappa u)^{-1}]$$
$$d = (\beta_{ss}^2 + \kappa_j)(\gamma - \delta\beta_{ss})$$

The model has a saddle-node bifurcation when $d = 0$ ($\beta_{ss} = \gamma/d$ as above). Hopf bifurcations are prescribed by the condition $bc = d$ with b, c and $d > 0$. Any Hopf points lie on the lower branch of solutions. According to the parameter values, there may be two, one or no such bifurcations. We will be particularly concerned with the first of these cases. The lower solution is unstable for some range of μ given by $\mu_2^* \leq \mu \leq \mu_1^*$, with $\mu_2^* > 0$ and $\mu_1^* < \gamma/\delta e$ as shown in figure 1.

4. Specific Cases.

4.1 Period-1 oscillations and influence of uncatalyzed reaction. $\gamma = 1.0, \delta = 0.1, \kappa_u = 7 \times 10^{-3}$.

The simplest dynamic response for this model is that of period-1 oscillations. Figure 2(a) shows the emergence of period-1 solution from the Hopf bifurcation points for the system with the above parameter values and the variation of amplitude across the oscillatory range. In all cases studied, the Hopf bifurcations are supercritical yielding a stable limit cycle. For the present case, the period-1 solution is stable for all values of μ between the Hopf points. There are two sections of the curve for which the amplitude varies rapidly with μ: that near the upper Hopf point will be of interest later.

If the uncatalyzed reaction rate constant is increased, the Hopf bifurcation points move closer together, decreasing the range of stationary-state instability. (In the isothermal autocatalator model, Hopf bifurcations exist only if $\kappa_u < 1/8$.) Figure 2(b) shows the variation in amplitude of the period-1 limit cycle solution for a system with $\kappa_u = 5 \times 10^{-2}$: as well as having small range, the "canard" behaviour noted above has been lost so the amplitude varies relatively smoothly with μ.

For smaller values of κ_u on the other hand the period-1 solution may bifurcate. With $\kappa_u = 5.5 \times 10^{-3}$ a stable period-2 solution emerges as a Floquet multiplier for the period-1 limit cycle passes through -1. The period-doubling bifurcation here is supercritical so the period-2 oscillation is stable. There is a second bifurcation at which the period-2 disappears and the period-1 regains stability (figures 2c and d).

If the uncatalyzed rate constant is decreased further, the character of the upper period-doubling bifurcation (that at larger μ) changes. for $\kappa_u = 10^{-3}$, this becomes a subcritical bifurcation, figure 2(e) and (f). An unstable period-2 solution emerges, growing in amplitude as μ increases. There is a turning point in the period-2 locus and the limit cycle is stable along the upper branch before disappearing at the lower period-doubling point. Such a system will thus show hard-excitation from period-1 into period-2 oscillations as μ is decreased.

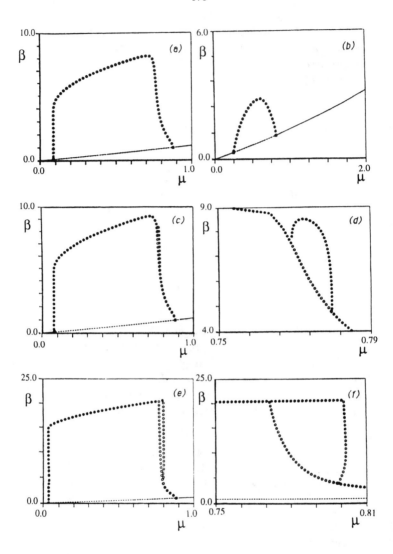

FIGURE 2. A series of bifurcation diagrams showing variation in amplitude of periodic solutions with μ, for various values of reaction rate of constant κ_u. These have been calculated using Auto. In all cases $\gamma = 0.1$: (a) $\kappa_u = 7.0 \times 10 - 3$, stable period one oscillation. (b) $\kappa_u = 5.0 \times 10 - 2$, stable period one oscillation with smaller maximum amplitude. (c) $\kappa_u = 5.5 \times 10 - 3$, the period one oscillation now bifurcates at a supercritical period doubling point. (d) $\kappa_u = 5.5 \times 10 - 3$, close-up of the period 2 region in (c). (e) $\kappa_u = 1.0 \times 10 - 3$, subcritical period doubling bifurcation. (f) close up of unstable period 2 oscillation shown in (e).

4.2 Influence of heat transfer rate : chaos. Returning to the system with $\delta = 0.1, \kappa_u = 5.5 \times 10^{-3}$ and $\gamma = 1.0$, we see Hopf bifurcation and period-doubling, as shown in figure 2(c) and (d). If the heat transfer parameter γ is decreased, corresponding to slower heat loss, further complexity is found. Between $\gamma = 0.7$ and 0.65, a second pair of period-doubling appear, as shown in figures 3(a) and (b). Now stable period-4 oscillations exist over a narrow range of the precursor concentration μ. Yet more period-doublings (bubbling) are rapidly introduced by further decreasing γ and by $\gamma = 0.5$ a full cascade [10] to chaotic responses has been achieved, figure 3(c).

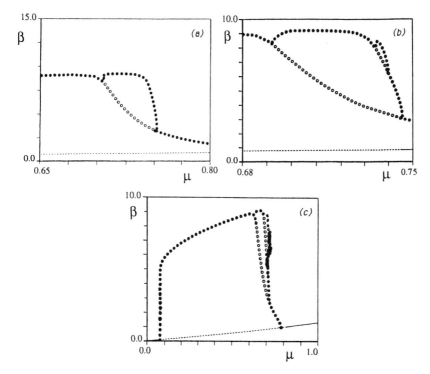

FIGURE 3. Series of period doubling bifurcations or 'bubbling', leading to chaois as the heat transfer coefficient γ is varied. In all cases $\kappa_u = 5.5 \times 10 - 3$, $\gamma = 0.1$. (a) $\gamma = 0.7$, a simple period 1, period 2, period 1 sequence. (b) By $\gamma = 0.65$ a period 4 solution has appeared. (c) By $\gamma = 0.5$ a full period doubling cascade takes place and region of chaotic solutions exists.

Representative time series (showing the autocatalyst concentration β as a function of time) are presented in figure 4. Considering a sequence with decreasing μ, following the upper Hopf point a period-1 limit cycle emerges. This period-doubles, and a typical period-2 trace for $\mu = 0.715$ is given in figure 4(j). By $\mu = 0.710$, a period-4 oscillation (figure 4(i)) is found. For $\mu = 0.708$, the time trace (figure 4(h)) is chaotic. Figure 5(a) shows the corresponding attractor in phase-space : the next amplitude map, figure 5(b), has a sharp peak. A window of period-3 responses emerges at smaller μ figure 4(f–g) and 5(c)) before another region of aperiodicity. In this lower chaotic range, figure 4(d–e) and 5(d), the next amplitude maps have much flatter maxima, as shown for $\mu = 0.690$ in figure 5(e). For $\mu = 0.68870$, period-4 has returned, figure 5(f) : later period-halvings occur as μ is decreased further.

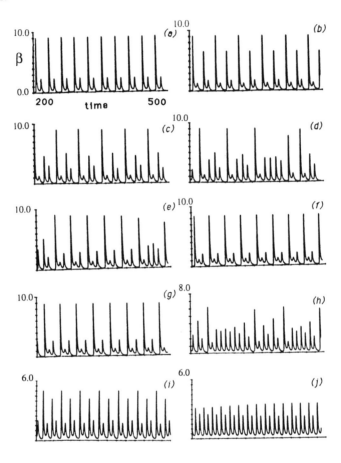

FIGURE 4. Time traces for $\gamma = 0.7$, $\gamma = 0.1$, $\kappa_u = 5.5 \times 10 - 3$, corresponding to the bifurcation diagram in figure 3(c). (a) $\mu = 0.65$, period 2. (b) $\mu = 0.687$, period 4. (c) $\mu = 0.6887$, period 4. (d) $\mu = 0.69$, chaotic trace. (e) $\mu = 0.695$, chaotic trace with period 3 solution emerging. (f) $\mu = 0.696$, period 3 window. (g) $\mu = 0.705$, period 3. (h) $\mu = 0.708$, smaller amplitude chaotic trace. (i) $\mu = 0.71$, period 4. (j) $\mu = 0.715$, period 2.

In these cases the attractors are folded almost at right angles, so as to lie approximately on the planes $\alpha = 0$ and $\theta = 0$ respectively. The fast motion corresponds to the sharp decrease in α and sharp pulse in β. The decay of the autocatalyst concentration is accompanied by a rapid increase in temperature. The system then cools exponentially, on a longer timescale through Newtonian heat transfer: during this period the concentration of A begins to increase again.

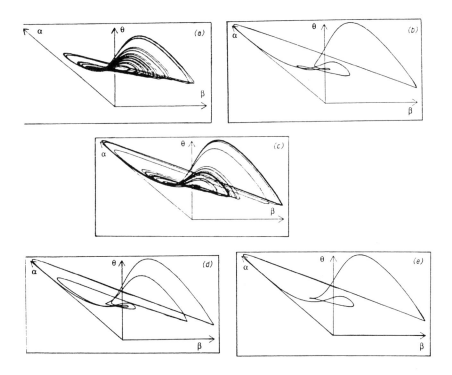

FIGURE 5. Three dimensional diagrams of the attractors and limit cycles corresponding to the time traces in figure 4. (a) $\mu = 0.69$, chaotic attractor showing characteristic folding. (b) $\mu = 0.696$, period 3 limit cycle. (c) $\mu = 0.708$, chaotic attractor. (d) $\mu = 0.71$, period 4 limit cycle. (e) $\mu = 0.715$, period 2 limit cycle.

4.3 Mixed-mode oscillations. The variation of the period-1 oscillation with μ for a system with $\kappa_u = 10^{-3}, \gamma = 0.05$ and $\delta = 0.025$ is shown in figure 7. This solution is stable, except for the range $0.5 < \mu < 0.56$. Again, we find a period-doubling sequence as μ is decreased (figure 8(m and n)). The time series then becomes chaotic figure 8(l). The corresponding next-amplitude map again has a sharp maximum as shown in figure 10(a). As μ is decreased from 0.5587 to 0.5586, figure 8(k), there is a significant change in the waveform, although this remains aperiodic. for the larger value of μ, the maximum value of β attained is ca. 8 : for $\mu \leq 0.5586$ larger amplitude excursions appear, with β attaining instantaneous values of ca. 20. We can now identify small, intermediate and large oscillatory pulses. This change in waveform sees a further sharpening of the next-maximum map, figure 10(b). Periodicity returns by $\mu = 0.5580$: the time series, figure 8(j), has a typical mixed-mode form with eight small peaks separating each large excursion, a 1^8 pattern. The amplitude of successive small peaks increases and there is then a period with $\beta \approx 0$ before the large peak.

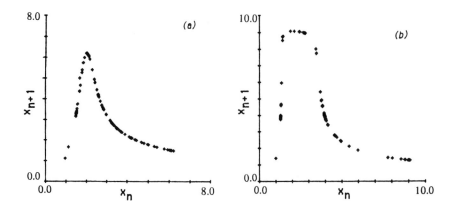

FIGURE 6. Net maxima plots of $x_{n+1} vs\ x_n$, for β which correspond to the chaotic attractors at $\gamma = 1.0$, $\delta = 0.1$ $k_\mu = 5.5 \times 10 - 3$, (a) $\mu = 0.69$, (b) $\mu = 0.708$.

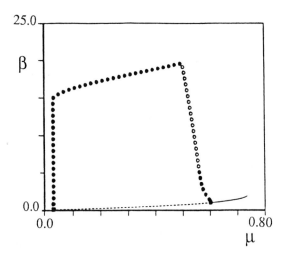

FIGURE 7. Bifurcation diagram showing the period 1 limit cycle at $\gamma = 0.05$, $\delta = 0.025$, $k_\mu = 1.0 \times 10 - 3$. The open circles show the unstable region where more complicated mixed mode oscillations arise. These are shown in figure 8.

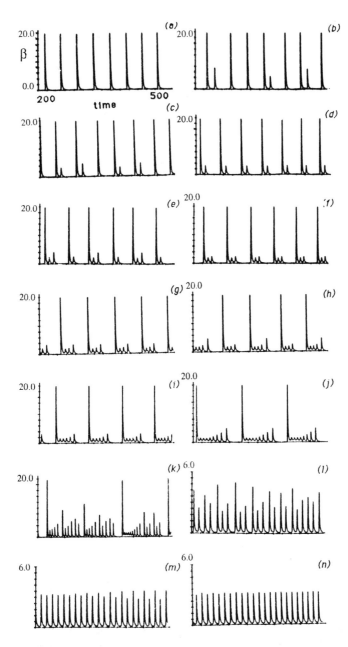

FIGURE 8. A sequence of mixed mode oscillations and small amplitude chaos for $\gamma =$ 0.05, $\delta = 0.025$, $k_\mu = 1.0 \times 10 - 3$. (a) $\mu = 0.498$, large amplitude period 1. (b) $\mu = 0.50$ $1^1 3^1$ pattern. (c) $\mu = 0.505$, $1^1 1^1 2^1$ pattern. (d) $\mu = 0.51$, $1^1 3^1$ pattern. (e) $\mu = 0.52$, $1^1 1^2$ pattern. (f) $\mu = 0.53$, 1^2 pattern. (g) $\mu = 0.535$, 1^3 pattern. (f) $\mu = 0.53$, 1^2 pattern. (g) $\mu = 0.535$, 1^3 pattern. (h) $\mu = 0.54$, $1^3 1^4$ pattern. (i) $\mu = 0.55$, 1^5 pattern. (j) $\mu = 0.5580$, 1^8 pattern. (k) $\mu = 0.5586$, large amplitude chaotic trace. (l) $\mu = 0.5587$, small amplitude chaotic trace. (m) $\mu = 0.5595$, period 2. (n) $\mu = 0.56$ period 1.

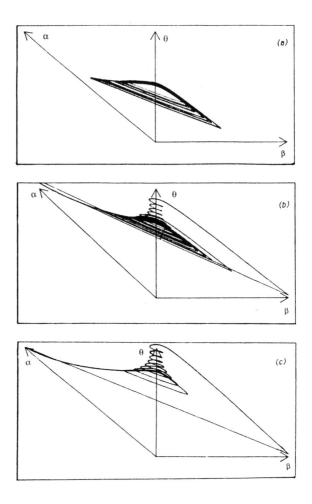

FIGURE 9. Chaotic attractors and mixed mode limit cycle at (a) $\mu = 0.5587$, (b) $\mu = 0.5586$, (c) $\mu = 0.5580$ in figure 8. The view chosen shows the escape from and spiral-shaped reinjection into the stable manifold.

The development of the attractor in phase-space through these changes of os-cillatory form can be seen in figure 9(a–c). for $\mu = 0.5587$, the trajectory winds around a small, relatively flat attractor appropriate to the small amplitude chaos seen in the time series. This part of the flow is also evidenced with $\mu = 0.5586$ in figure 9(b). Now the winding also gives way to an occasional large excursion. During this, the autocatalyst concentration falls virtually to zero : the temperature excess then falls, exponentially, whilst the concentration of A increases. At some point there is then a burst of conversion of A to B with a following temperature rise. The trajectory returns to the original portion of the attractor by spiralling in from above. The motion on this attractor gives large amplitude chaos. With $\mu = 0.5580$, figure 9(c), the original structure has resolved itself into a true limit cycle with eight turns between the onset of the large amplitude excursion and the re-injection.

The number of small peaks in the cycle decreases as μ is made smaller : there is a 1^4 response at $\mu = 0.550$ and a 1^3 at $\mu = 0.535$. In between such parent forms there are higher-order concatenations, e.g. with $\mu = 0.5400$ we find a $1^4 1^3$ state, figure 8(h). Similarly, there is a $1^2 1^1$ state at $\mu = 0.520$ (figure 8(e)) between the 1^2 at $\mu = 0.530$ and the 1^1 at $\mu = 0.510$ (figures 8(f) and (d) respectively). Amongst other waveforms found between the 1^1 and the simple period-1 (1^0) are a $(1^1)^2 1^2$ at $\mu = 0.505$ (figure 8(c)) and a $1^1 (1^0)^2 1^1 1^0$ (or $3^1 2^1$) at $\mu = 0.500$ (figure 8(d)).

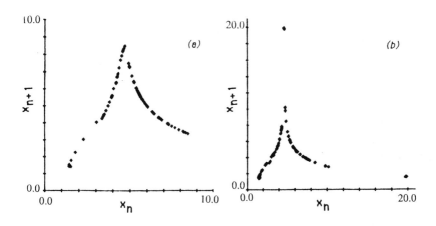

FIGURE 10. Next maxima plots of x_{n+1} vs x_n, for β which correspond to the chaotic attractors at $\gamma = 0.05$, $\delta = 0.25$, $\kappa_u = 1.0 \times 10 - 3$, (a) $\mu = 0.5587$, (b) $\mu = 0.5586$.

5. Discussion. The model under consideration here comprises the interaction of chemical and thermal feedback. The exothermicity of the reaction is indicated through the parameter δ whilst γ is a measure of the heat transfer coefficient for the system. With the particualr parameter values studies in this paper, we are concerned mainly with reactions having only moderate exothermicities ($\delta = 0.1$ or 0.025). Thus, much of the evolution is virtually isothermal. Only when chemical autocatalysis produced a high concentration of the species B does the heat release rate become significant. The sharp rise of the autocatalyst concentration occurs on a very short timescale : during this short period there can be little heat transfer so the system departs from isothermal operation.

The actual temperature excursions can be evaluated. If the temperature of the reacting mixture is T and the ambient temperature T_a, the temperature excess $\Delta T = T - T_a$ is given by

$$\Delta T = \theta(RT_a^2/E).$$

Taking the results from figure 7, we see that with $\delta = 0.025, \theta <$ ca. 05. If we assume an activation energy of 100 $KJ \ mol^1$ and an ambient temperature of 298K, we have $\Delta T <$ ca. 4K. For the time series in figure 8(1), the temperature excess will not exceed 2K. Clearly, then large non-isothermal effects are not required to produce even the highest degrees of complexity in this scheme.

The origins of the internal forcing within the present model can be seen relatively clearly. The autonomous oscillations in the concentration of B, which occur even with $\delta = 0$, lead to oscillations in the temperature. This then provides a forcing to the chemical kinetics through the term μe^θ. If the argument θ does become large we could approximate the exponential dependence with the expansion $\mu(1 + \theta + \dots)$. As θ is an oscillatory function this form provides a similar forcing to that obtained by, say, a sinusoidal non-autonomous external forcing such as $\mu(1 + A\sin(\omega\tau))$.

The evolution of mixed-mode oscillations and their progression to chaotic behaviour in section 4.3 above shows some interesting features. If we consider here the traverse through parameter space with increasing μ we find that the bifurcation from period-1 to mixed-mode at $\mu = $ ca.05 is accompanied by a Floquet multiplier for the limit cycle leaving the unit circle rapidly through -1. We have not found evidence for complex Floquet multipliers in this system so there is no quasiperiodicity or bifurcation of the limit cycle to a torus. Recent discussions of the relationships between mixed-mode waveforms and toroidal behaviour have been given by Barkley [5,6], Schell [21] and others [11]. Maselko & Swinney [16–18] have observed periodic states similar to that shown in figure 8(b) with $\mu = 0.5$, which they have interpreted as frequency-locked motion on a four-torus. In our system there are, however, only three variables.

The spiralling nature of the mixed-mode attractor can be related to the eigenvalues of the stationary-state which are constituted by a weakly attracting eigenvalue λ_1 and a complex pair with positive real parts (these real parts of higher magnitude than λ_1). At slightly higher values of the reactant concentration, there appears to be a homoclinic tangency, but because of the relative magnitudes of the attracting and repelling eigenvalues, this must be of a different form from the familiar manifestation of Shilnikov homoclinicity [12] such as that found in the Lorenz system [24]. The sequence of mixed-mode $\rightarrow$ large amplitude chaos $\rightarrow$ small amplitude chaos $\rightarrow$ small amplitude oscillations does not appear to be characterized by any of the current scenarios for the development of chaos, but has been reported recently by Albahadily et al. [1] in a different (experimental) situation.

List of Symbols

A	intermediate in reaction scheme
a	concentration of species A
A_0	pre-exponential factor in Arrhenius temperature dependence
B	autocatalytic intermediate
b	concentration of species B
c_p	molar heat capacity of mixture
c_0	molar density of mixture
C	final product
E	activation energy
k_i	rate constant of i-th step
P	precursor reactant
p	concentration of P
Q	exothermicity of reaction
S	surface area across which heat is transferred
T	temperature of reacting mixture
T_a	ambient temperature of surroundings
V	reactor volume
α	$= (k_1/k_2)^{1/2}a$, dimensionless concentration of A
β	$= (k_1/k_2)^{1/2}b$, dimensionless concentration of B
γ	$= \chi^{S/k}2^c c_0 V$, dimensionless heat transfer coefficient
δ	$= Q(k_2/k_1)^{1/2}E/c_p c_0 RT_a^2$, dimensionless exothermicity
ε	$= k_0/k_2$, dimensionless precursor decay rate constant
θ	$= (T - T_a)E/RT_a^2$, dimensionless temperature rise
κ_u	$= k_3/k_2$, dimensionless uncatalyzed reaction rate constant
μ	$= k_0 k_1^{1/2} p_0/k_2^{3/2}$, dimensionless precursor concentration
χ	surface heat transfer coefficient

Acknowledgements. We are grateful to Professor K. Showalter and Mr. Bo Peng (University of West Virginia) for helpful discussion; SKS thanks NATO (grant no. 0124189) and the Institute for Mathematics and its Applications, University of Minnesota for financial support.

REFERENCES

[1]   F.N. ALBAHADILY, J. RINGLAND AND M. SCHELL, J. Chem. Phys., 90 (1989), pp. 813–821.

[2]   F. ARGOUL, A. ARNEODO, P. RICHETTI AND J.C. ROUX, *From quasiperiodicity to chaos in the Belousov–Zhabotinskii reaction I and II*, J. Chem. Phys., 86 (1987), 3325–3356.

[3]   F. ARGOUL, A. ARNEODO, P. RICHETTI AND J.C. ROUX AND H.L. SWINNEY, *Chemical chaos : from hints to confirmation*, Acc. Chem Res., 20 (1987) 436–442.

[4]   K. BAR–ELI AND R.M. NOYES, *Computations simulating experimental observations of complex bursting patterns in the Belousov–Zhabotinskii system*, J. Chem. Phys., 88 (1988), 3646–3654.

[5]   D. BARKLEY, *Near critical behavior for one-parameter families of circle maps*, Phys. Lett., A 129 (1988), 219–223.

[6]   ——————, *Slow manifolds and mixed-mode oscillations in the Belousov–Zhabotinskii reaction*, J. Chem. Phys., 89 (1988), 5547–5599.

[7]   D. BARKLEY, J. RINGLAND AND J. TURNER, *Observations of a torus in a model of the Belousov–Zhabotinskii reaction*, J. Chem. Phys., 87 (1987), 3812–3820.

[8]   P. BERGE, Y. POMEAU AND C. VIDAL, *Order within chaos*, Wiley, New York, 1984.

[9]   E. DOEDEL, *AUTO : continuation and bifurcation problems in ordinary differential equations* (1986).

[10] M.J. FEIGENBAUM, *Universal behavior in nonlinear systems*, Los Alamos Sci., 1 (1980), 4–27.

[11] V. FRANCESCHINI, *Bifurcations of tori and phase-locking in a dissipative system of differential equations*, Physica, 6D (1983), 285–304.

[12] J. GUCKENHEIMER AND P. HOLMES, *Nonlinear oscillations, dynamical systems and bifurcations of vector fields*, Springer, New York, 1983.

[13] J.L. HUDSON, M. HART AND D. MARINKO, *An experimental study of multiple peak periodic and nonperiodic oscillations in the Belousov–Zhabotinskii reaction*, J. Chem. Phys., 71 (1979), 1601–1606.

[14] J.L. HUDSON AND J.C. MANKIN, *Chaos in the Belousov–Zhabotinskii reaction*, J. Chem. Phys., 74 (1981) 6171–6177.

[15] R.D. JANZ, D.J. VANACEK AND R.J. FIELD, *Composite double oscillation in a modified version of the Oregonator model of the Belousov–Zhabotinskii reaction*, J. Chem. Phys., 73 (1980), 3132–3138.

[16] J. MASELKO, *Experimental studies of complicated oscillations*, Chem. Phys., 51 (1980), 473–480.

[17] J. MASELKO AND H.L. SWINNEY, *Complex periodic oscillations and Farey arithmetic in the Belousov–Zhabotinskii reaction*, J. Chem. Phys., 85 (1986), 6430–6441.

[18] ——————, *Phys. Lett.*, A 119 (1987), 403–406.

[19] J.H. MERKIN, D.J. NEEDHAM AND S.K. SCOTT, Proc. R. Soc. Lond., A 406 (1986), 299–323.

[20] O.E. RÖSSLER, *An equation for continuous chaos*, Phys. Lett., A 57 (1976), 397–398.

[21] M. SCHELL, AND F. ALBAHADILY, J. Chem. Phys., 90 (1989) 821–829.

[22] R.A. SCHMITZ, K.R. GRAZIANI AND J.L. HUDSON, *Experimental evidence of chaotic states in the Belousov–Zhabotinskii reaction*, J. Chem. Phys., 67 (1977), 3040–3044.

[23] K. SHOWALTER, R.M. NOYES AND K. BAR–ELI, *A modified Oregonator model exhibiting complicated limit cycle behavior in a flow system*, J. Chem. Phys., 69 (1978), 2514–2524.

[24] C. SPARROW, *The Lorenz equations : bifurcations, chaos and strange attractors*, Springer, New York, 1982.

BIFURCATIONS AND GLOBAL STABILITY IN SURFACE CATALYZED REACTIONS USING THE MONTE CARLO METHOD*

D.G. VLACHOS, L.D. SCHMIDT AND R. ARIS**

Abstract. The vector of state variables characterizing the dynamics of heterogeneous systems is often spatially nonuniform. Ordinary differential equations (ODEs) or partial differential equations (PDEs) can describe spatial and temporal evolution of nonuniform field systems only under severe assumptions. The Monte Carlo method (MCM) is applied to model such heterogeneous systems. The conditions under which ODEs or PDEs fail are examined by studying generic model systems using the stochastic method. In the presence of nonlinearities in the governing deterministic equations, spatial inhomogeneities are observed. Multiplicities, cusp points, and periodic solutions are calculated by systematic investigation in parameter space and the global stability of solutions is presented. It is demonstrated that the presence of imperfections on surfaces introduces local nonuniformities in the adlayer (nucleation centers) and their interaction on the synchronization of the surface is discussed.

1. Introduction.

Studies of heterogeneous reactions in many applications such as catalysis, electrochemistry, and chemical vapor deposition often reveal complicated behavior: multiplicities in reaction rates, phase transitions, oscillations, and propagating or standing waves [1-7]. The dynamics and interactions occurring at the microscopic level can result in exotic macroscopic patterns varying from ordered structures, to disordered structures, to turbulence so far associated only with fluids [1,7]. The macroscopic structure developed by the microscopic interactions is related to the catalyst activity and the overall performance of the system. Therefore, an understanding of the macroscopic patterns formed at atomic level is required along with the performance obtained.

Most models consider spatially averaged concentrations of adsorbed species on the catalyst surface by solving a set of ODEs [2,5,6]. However, the macroscopic patterns developed during surface phenomena usually correspond to a nonuniform distribution of species on the surface [8,9]. Hence, a spatially independent concentration field is in many cases an oversimplification of reality. Spatial inhomogeneity of the surface controls the bifurcation behavior which therefore cannot be modeled using ODEs [8]. The role of local fluctuations in concentration is also ignored by ODE models, and its influence on the dynamics of the system is missing. An elucidation of the mechanism for kinetic oscillations requires an investigation of all processes affecting the reaction rate at a molecular level. This can be achieved by using an atomic level model, such as the Monte Carlo method (MCM) [10].

Recently, the MCM was used to study the phase transitions associated with a bimolecular reaction between A and B_2 (one monatomic and one diatomic species)

---

*This research partly supported by NSF under grant No. CBT 9000117, a Graduate School Fellowship, the Minnesota Supercomputer Institute, and the Army High Performance Computing Research Center.

**Department of Chemical Engineering and Materials Science, University of Minnesota, Minneapolis, MN 55455.

or A and B (two monatomic species) [9,11-13]. We also reported phase transitions, multiplicities, and oscillations for a unimolecular reaction in a continuous stirred flow reactor [8,14]. In our studies, the roles of surface diffusion and defects on phase transitions and transient path of the system were examined.

In this paper, an outline of the deterministic theory used to describe the dynamics of heterogeneous systems is first presented. Then, the application of the MCM to surface science problems is explained. The advantages and disadvantages of utilizing the MCM to model the bifurcation behavior and stability of solutions in heterogeneous systems are discussed. Deviations between the deterministic and the stochastic model are reported, and the parameters affecting the deviations are identified. This is achieved by simulation of two model systems: unimolecular and bimolecular reactions. The role of spatial inhomogeneities introduced by absorbate interactions and the presence of surface imperfections on the multiplicity and stability of the solutions is demonstrated. The relation between fluctuations and oscillations is examined, and regimes of parameter space where fluctuations become important are discussed.

2. Mean Field Theory for Nonlinear Dynamics in Surface Reactions.

2.1 Spatially homogeneous systems. Since early 1960's, the common approach in modeling the dynamics of surface reactions is to use a lumped-parameter system described by a set of ODEs

$$(1) \qquad \frac{d\mathbf{y}}{dt} = \mathbf{F}(\mathbf{y}, \mathbf{p})$$

which is known as mean field (MF) or Bragg-Williams approximation. Here, $\mathbf{F}$ is the vector field (assumed to be differentiable) describing the kinetics in the adlayer, $\mathbf{y}$ is a vector of n state variables (phase space) such as concentrations of species involved in the surface reaction, gas pressure, and/or surface temperature, and $\mathbf{p}$ is a vector of parameters in m-dimensional parameter space. Most surface reaction systems involve many parameters, but usually, only a few of them are varied over a wide range. Thus, the dimension m of parameter space is usually 2 or 3.

Insight into the importance of the interplay between the spatial structure and the dynamics of the system as well as comparison of the MF theory with the MCM can be gained by modeling generic model systems. The first model system examined (presented here in more detail) is a unimolecular reaction in a continuous stirred tank reactor (CSTR). The dynamics in that case are governed by the equations [5,14]

$$(2) \qquad \frac{d\theta}{dt} = r_a - r_d - r_R$$

for the surface concentration θ, and

$$(3) \qquad \frac{dP_A}{dt} = \frac{P_{A,o} - P_A}{\tau_R} + P^*(r_d - r_a)$$

for the partial pressure P_A of reactant A. Here, τ_R is the residence time in the reactor, $P_{A,o}$ is the pressure of reactant at the inlet, P^* is a parameter expressing the ratio of catalyst surface to reactor volume (surface capacity), and r_a, r_d, r_R are the adsorption, desorption, and reaction rates.

In the presence of interactions of strength w (in our formulation, attractive interactions imply $w > 0$), Eq. (2) is written

$$(4) \qquad \frac{d\theta}{dt} = k_a P_A(1 - \theta) - k_d \exp(-z_s\theta w/kT)\theta - k_R\theta,$$

where z_s is the number of neighbors of each surface site, k is the Boltzmann constant, T is the temperature, and k_a, k_d, k_R are the adsorption, desorption, and reaction rate constants. In this model system, the effect of the exponential nonlinear desorption term on the inhomogeneity of vector $\mathbf{y}$ and the consequences on the kinetics are investigated.

The second model system represents a bimolecular reaction between two species A and B [11,12]

$$(5) \qquad \frac{d\theta_A}{dt} = k_{aA} P_A(1 - \theta_A - \theta_B) - k_{dA}\theta_A - k_R\theta_A\theta_B$$

$$(6) \qquad \frac{d\theta_B}{dt} = k_{aB} P_B(1 - \theta_A - \theta_B) - k_{dB}\theta_B - k_R\theta_A\theta_B$$

(known as Langmuir-Hinshelwood kinetics) where the subscripts denote the corresponding species. In this system (hereafter called AB model), the partial pressures in the gas phase are assumed to be uniform in space and constant in time. More complicated kinetics can be appropriate for desorption (as in the first model) or reaction. However, deviations between the deterministic (outlined above) and stochastic model can be observed caused by the simple nonlinear reaction term appearing in Eqs. (5) and (6).

2.2 Bifurcation analysis and stability. As steady state, the right hand side of Eq. (1) will be equal to zero. The stationary solutions (fixed points) of

$$(7) \qquad\qquad\qquad \mathbf{F}(\mathbf{y},\mathbf{p}) = 0$$

are classified based on the eigenvalues of the linearized Jacobian around the fixed point

$$(8) \qquad\qquad J = \begin{bmatrix} \dfrac{\partial F_1}{\partial y_1} & \dfrac{\partial F_1}{\partial y_2} \\[2mm] \dfrac{\partial F_2}{\partial y_1} & \dfrac{\partial F_2}{\partial y_2} \end{bmatrix}_{SS}$$

and the stability is determined by the trace and the determinant of the Jacobian [5]

$$(9) \qquad\qquad\qquad \det J > 0$$

$$(10) \qquad\qquad\qquad \operatorname{tr} J < 0.$$

In a two-dimensional phase space, there are five kinds of fixed points: turning point (zero eigenvalue), stable or unstable point (real non-zero eigenvalue), and stable or unstable foci (complex conjugate eigenvalues). In addition, limit cycles describing periodic solutions (closed curves in phase space) are frequently found. Complicated nonstationary solutions are often observed ranging from regular to irregular to chaos [1].

It is often possible to express the $n - 1$ elements of vector $\mathbf{y}$ as an explicit or implicit function of the nth element. This provides $n - 1$ relations which can be used so that the steady state is described by a single equation, $f(y, \mathbf{p}) = 0$. The steady state manifold in $(y, \mathbf{p})$ space can be defined as a set of all points $(y, \mathbf{p})$ satisfying this equation. A singular point $(y^o, \mathbf{p}^o)$ is said to be of codimension k if it satisfies

$$(11) \qquad f(y^o, \mathbf{p}^o) = \frac{df}{dy}(y^o \mathbf{p}^o) = \cdots = \frac{d^k f}{dy^k}(y^o, \mathbf{p}^o) = 0$$

$$(12) \qquad \frac{d^{k+1} f}{dy^{k+1}}(y^o, \mathbf{p}^o) \neq 0.$$

Singular points of codimension 1 and 2, known as fold and cusp points respectively, will be discussed next. Considering one parameter of the m dimensional vector $\mathbf{p}$ as the bifurcation parameter λ, the steady state equation is written

$$(13) \qquad f(y, \lambda; \mathbf{p}) = 0,$$

and a bifurcation diagram of y vs. λ can be generated and compared with experimental data.

For example, construction of a single equation for the unimolecular reaction model (Eqs. (3) and (4)) results in the following equation

$$(14) \qquad k_a(P_{A,o} - \tau_R k_R P^* \theta)(1 - \theta) - k_d \exp(-z_s \theta w/kT)\theta - k_R \theta = 0$$

In a differential reactor ($\tau_R \to 0$), the above equation becomes (hereafter called isotherm)

$$(15) \qquad KP_A = \frac{\theta[k_R/k_d + \exp(-z_s \theta w/kT)]}{1 - \theta}$$

where $K = k_a/k_d$ is the equilibrium adsorption-desorption constant.

2.3 Spatially inhomogeneous systems. Application of Eq. (1) requires that all variables are uniform along the surface, i.e. there exists an infinite diffusion coefficient of Fickian type so that perfect "mixing" is accomplished in the adsorbed layer. This can be achieved in a homogeneous phase, as for example the Belousov-Zhabotinsky reaction in a liquid phase [15]. However, it seems very difficulty to control the "stirring" on a catalyst surface (heterogeneous system). As a consequence, most experimental and industrial heterogeneous catalytic systems are innately distributed, with concentration and/or temperature gradients. Spatial inhomogeneities are usually caused by interactions between the adatoms, surface imperfections (dislocations, grain boundaries, point defects), nonuniform distribution of impurities, or different crystallographic planes present (as in supported catalysts). The vector **F** is a set of nonlinear functions owing to the cooperative phenomenon of interactions, inhibition, or autocatalysis [2,5,6]. This phenomenon determines the spatial nonuniformities on the surface.

Despite the nonuniformity of vector field **y** along the surface, very few studies have focused on spatial characteristics of oscillations and steady states. More recent studies include the thermographic imaging performed by Schmitz et al. [16,17], the in situ scanning LEED and the photoemission electron microscopy [6,18,19], and the photodiode monitor of temperature in methylamine decomposition on electrically heated wires by Cordonier et al. [20].

Distributed systems in surface reaction systems can be described by PDEs of the form

$$(16) \qquad \frac{\partial y}{\partial t} = -\nabla \cdot (-\mathbf{D} \cdot \mathbf{y}) + \mathbf{F}(\mathbf{y}, \mathbf{p})$$

where $-\mathbf{D} \cdot \nabla \mathbf{y}$ is the diffusive flux and **D** is the diffusion tensor. Eq. (16) is a generalization of the well known second Fick's law for an open system.

Most models explaining oscillations in catalytic systems assume either explicitly or implicitly some type of attractive interactions. In the presence of attractive interactions, there is a net flux from a low concentration (single atoms) towards a high concentration (clusters) caused by minimization of Gibb's free energy. In contrast, Eq. (16), which uses a Fickian diffusion, predicts a flux from a high to a low concentration. This implies that Fickian diffusion, as that described by Eq. (16), tends to homogenize the distribution of species and in the limit of an infinite Fickian diffusion, Eq. (1) would be more accurate. Therefore, Eq. (16) is structurally incorrect in the presence of attractive interactions [21] and cannot predict the spatial and temporal evolution of catalytic processes. In the following section, the modeling of spatial and temporal patterns in heterogeneous systems using the MCM is outlined.

3. The Monte Carlo Method.

3.1 The Monte Carlo method as a stochastic simulation model. The MCM was developed by von Neumann, Ulam, and Metropolis at the end of the Second World War at Los Alamos to study the diffusion of neutrons in fissionable material. The name "Monte Carlo", chosen because of the extensive use of random numbers in the simulation, was coined by Metropolis in 1947 and used in the title of a paper describing the early work at Los Alamos [22]. However, the history of MCM is even longer. In 1901, the Italian mathematician Lazzerini performed a simulation by spinning round and dropping a needle 3407 times. He estimated π to be 3.1415929. In the beginning of the century, statisticians had also used model sampling experiments to investigate some problems (the correlation coefficients in the "t" distribution, trajectories to study the elastic collisions with shaped walls). The novel contribution of von Neumann, Ulam, and Metropolis was to realize that determinate mathematical problems could be treated by finding a probabilistic analogue which was then solved by a stochastic sampling experiment [10].

The MCM is appropriate for study of phenomena in heterogeneous systems because these phenomena are discrete and microscopic. MC simulations play a valuable role in providing essentially exact results for problems in surface science which would be otherwise only soluble by approximate methods or might be quite intractable. In this sense, MC computer simulation is a test of deterministic theory. MC simulations provide a direct route from the microscopic details of a system to macroscopic properties of experimental interest (equation of state, transport coefficients, structural order parameters, reactivity, and so on). It may be difficult or impossible to carry out experiments under extremes of temperatures and pressure, while a computer simulation would be perfectly feasible. Even though quite subtle details of molecular motion and structure in heterogeneous catalysis are difficult to probe experimentally, they can be extracted readily from a computer simulation. Simulations are performed under well defined conditions so that the effect of different microscopic mechanisms on the macroscopic behavior of the system can be elucidated. Finally, while the speed of molecular events is itself an experimental difficulty, it presents no hindrance to computer simulation.

In the MCM, various states of the system are generated using random numbers and weighted with appropriate probabilities. A macroscopic specimen of matter is simulated by means of a model system containing a relatively small number of atoms occupying elements of an array. To simulate a large system, periodic boundary conditions are used. The value of matrix elements indicate the type of species present at this element, if any. Thus the surface can be thought as a sparce matrix where the degree of sparcity shows the portion of the surface which is unoccupied. Each time an element of the matrix is randomly chosen and various probabilities are assigned to each matrix element based on which processes are simulated. At each point of parameter space, application of the MCM provides the degree of sparcity which is the average concentration of each species as well as the structure of the matrix, i.e. the distribution of species in space. A record of the matrix evolution gives the spatial and temporal variation of the state variables.

The limitations of the technique include the finite number of configurations averaged for the calculation of ensemble averages, the finite size of simulation cell, the boundary effects, and the pseudo-randomness of the random numbers used generated from multiplicative congruential generators. This procedure of random numbers generation is deterministic and periodic. Since the periodicity is usually large enough in comparison with the actual Markov chain length, no serious practical limitation arises. It is an inherent feature of systems simulated that they contain relaxation modes with very low frequencies. Such slow relaxation occurs in all models which exhibit conservation laws (Hydrodynamic slowing down), second order phase transitions (critical slowing down), first order phase transitions (metastable states), and at low temperatures compared to energetic barriers [10]. Then the approach towards equilibrium is extremely computer time consuming.

3.2 Application of the Monte Carlo method to surface catalyzed reactions. The MCM in statistical physics models equilibrium and nonequilibrium systems by stochastic computer simulation. Starting from a description of the desired physical system in terms of a model Hamiltonian, one uses random numbers to construct the appropriate probability with which the various generated states of the system have to be weighted. Studies involving conservation of particles are performed in a closed system (canonical ensemble). However, kinetics of heterogeneous systems are usually studied in an open system (grand canonical ensemble) and these will be discussed briefly below.

To set up the kinetics, a probability function $p(\mathbf{k})$ is introduced which gives the probability that a given microscopic configuration $\mathbf{k} = (k_1, k_2, \ldots, k_{N_s})$ occurs at time t, where N_s is the total number of adsorption sites on the surface (matrix elements). Since all relevant processes such as adsorption, desorption, diffusion, etc. are Markovian, the probability function satisfies a master equation

$$(17) \qquad \frac{dP(\mathbf{k})}{dt} = \sum_{\mathbf{k}'} W(\mathbf{k}' \to \mathbf{k})P(\mathbf{k}') - \sum_{\mathbf{k}} W(\mathbf{k} \to \mathbf{k}')P(\mathbf{k})$$

where $W(\mathbf{k}' \to \mathbf{k})$ is the transition probability per unit time that the microscopic state $\mathbf{k}'$ changes into the microscopic configuration $\mathbf{k}$, and the summation is over all possible states. The first summation is the rate of generation, and the second summation is the rate of consumption of state $\mathbf{k}$. W must satisfy the microscopic reversibility (or detailed balance condition)

$$(18) \qquad W(\mathbf{k}' \to \mathbf{k})P_o(\mathbf{k}') = W(\mathbf{k} \to \mathbf{k}')P_o(\mathbf{k})$$

where P_o is the equilibrium probability. Next, adsorption and desorption processes are considered (grand canonical ensemble). The equilibrium probability is given by [23]

$$(19) \qquad P_o(\mathbf{k}) = \exp(-H(\mathbf{k})/kT)/\Xi.$$

Here, H is the Hamiltonian for a grand canonical ensemble given by

$$(20) \qquad H(\mathbf{k}) = E(\mathbf{k}) - N(\mathbf{k})\mu$$

and Ξ is the partition function

$$(21) \qquad \Xi = \sum_{\mathbf{k},N} \exp[(-E(\mathbf{k}) + N\mu)/kT]$$

where $E(\mathbf{k})$ is the internal energy of state $\mathbf{k}$, $N(\mathbf{k})$ is the number of particles in the system, and μ is the chemical potential of the gas phase which is related to the gas pressure.

Using Eqs. (19)–(21), Eq. (18) becomes

$$(22) \qquad \frac{W(\mathbf{k}' \to \mathbf{k})}{W(\mathbf{k} \to \mathbf{k}')} = \exp[-(E(\mathbf{k}) - E(\mathbf{k}'))/kT] \exp[(N(\mathbf{k}) - N(\mathbf{k}'))\mu/kT].$$

Similarly, diffusion is an equilibrium process, and the principle of microscopic reversibility has to be invoked.

The final relation (22) is not sufficient to determine all transition probabilities. A further choice has to be made based on the physical system simulated. This has no effect as far as equilibrium (stationary solutions) are determined because of the Onsager axiom that equilibrium is independent of the route approached. Specifying the transition probabilities suitably, the MC averages converge to the exact value in the limit of infinite Markov chain length, since the method is ergotic. The probabilities and the algorithm used in this study are summarized in figure 1.

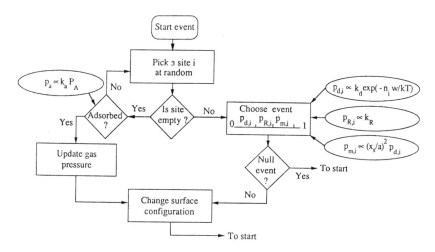

Figure 1. Flow chart of the MC algorithm and the probabilities of various events used in the unimolecular model system. n_i is the number of nearest neighbors surrounding atom at ith element and x_s/a is the mean number of atomic jumps during the residence time of an atom on the surface. Probabilities in the bimolecular reaction model system are defined in a similar way.

Even for the simple adsorption-desorption model (Eq. (22)), there is no analytical solution except at the critical point [23]. Systems including irreversible chemical reaction (nonequilibrium systems) are outside classical statistical mechanics, and no rigorous theories exist to describe them [8,9]. Since microscopic details of surface and spatial inhomogeneities are inherently included in the Monte Carlo method, it is a very suitable technique for these applications.

4. Comparison of the Results Obtained by the Monte Carlo Method and the Ordinary Differential Equations.

In this section, results obtained by the MCM are compared with the predictions of deterministic equations. In the simulations presented below, surface diffusion is not included. The pressure in the gas phase is first assumed constant in time (a differential reactor) and the solution of Eq. (4) is examined in parameter space. In order to investigate periodic solutions, the system of Eqs. (3), (4) is then modeled.

4.1 Spatial structure and pattern formation. Figure 2a shows a snapshot of the surface configuration obtained after 310^8 attempts in the presence of attractive absorbate-absorbate interactions ($w > 0$) for a certain point in parameter space. Agglomeration of adatoms takes place and clusters are formed at several positions of the surface caused by the dissipative structure of the system. Small clusters usually shrink and new clusters are formed at different locations. The effect of local fluctuations in adsorption on the formation of clusters and therefore on the spatially nonuniform solution is very profound. The role of cluster formation and growth on the stability of solutions and existence of oscillations will be discussed in the following sections.

In the presence of repulsive interactions ($w < 0$), atoms repel each other. When only first nearest neighbors are considered, and the surface is almost half filled, atoms occupy the diagonals of a square, as shown in figure 2b. Thus ordering on the surface occurs which results in a lower Gibbs free energy. A disordering-ordering transition (from a dilute gas to a ordered structure) can take place as a bifurcation parameter (pressure) changes. More complicated structures and transitions can occur on the surface if interactions of longer range are taken into account. The patterns produced depend on the range, strength, and type of interactions.

In addition to the adsorption-desorption mechanism examined, surface diffusion is expected to be another mechanism which also drives atoms towards the clusters minimizing the free energy [8]. Thus, in the presence of attractive interactions, "stirring" of the adlayer by diffusion does not seem very likely. Hence, the above examples indicate that the uniform distribution of atoms assumed in Eq. (1) is only a crude approximation, and spatially nonuniform solutions govern the dynamics of heterogeneous systems in the presence of absorbate interactions.

(a)

(b)

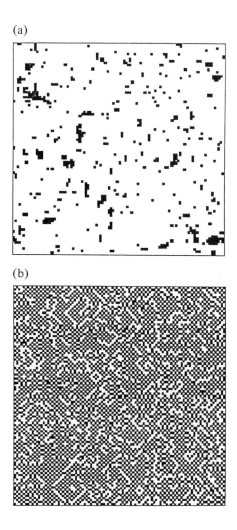

Figure 2. Snapshots of nonuniform solutions obtained by the MCM. Panel (a) shows a snapshot obtained from a 100×100 surface after 310^8 trials in the presence of attractive interactions ($w/kT = 2, k_R = 0, \theta_o = 0, KP_A = 0.0192$). Panel (b) shows a snapshot obtained from a 100×100 surface after 410^8 trials in the presence of repulsive interactions ($w/kT = -4, k_R = 0, \theta_o = 1, KP_A = 125$).

4.2 Multiple solutions and stability. In the limit of no reaction ($k_R/k_d = 0$), Eq. (4) has multiple solutions (three solutions) if the temperature is lower than a critical value ($w/kT = 1$) determined by the cusp point condition (Eqs. (11), (12)). Application of the equal area Maxwell rule determines the thermodynamic equilibrium between the two phases (tie line). States on the lower and upper branches are stable with respect to perturbations, i.e. the system is trapped in one of them for an infinite time. States in the intermediate branch are unstable with respect to perturbations, and trajectories lead either to the upper or lower branch depending on the direction of the perturbation.

Simulations using the MCM show that as the partial pressure of reactant A increases, the coverage increases (nonuniform solution) until a discontinuous (first order) phase transition from the low coverage branch to the high coverage branch is observed. This behavior is depicted in figure 3a. Time series of coverage shown in figure 3b, indicate that the system is trapped in the low state branch for a long time (metastable state) before the transition to the upper branch (stable state) take place. Growth of one or more big clusters developed on the surface by local fluctuations results in a phase transition and in a minimization of free energy.

This transition occurs above a critical value of pressure, $P_{A,c}$, and it is observed up to a maximum pressure, $P_{A,t}$. Simulations at all intermediate pressures exhibit a phase transition from the lower branch to the upper branch provided a sufficient long time. The closer the pressure is to $P_{A,c}$, the longer the metastable lives. At the critical point, $P_{A,c}$, an infinite time is required for the transition of the system. The last point at which the transition occurs, $P_{A,t}$, (hereafter called pseudospinodal point) corresponds to the turning point of the deterministic MF model. Further increase of the pressure results in only one stable solution without any metastable state.

Starting from a high reactant pressure (upper branch) and decreasing the pressure, a unique solution is obtained if the pressure is higher than $P_{A,c}$. Below this value, a phase transition from the upper branch (metastable state) to the lower branch (stable state) is observed after sufficient time as shown in figure 3b. Thus the isotherm exhibits hysteresis, and multiple solutions are predicted for an isothermal system as shown in figure 3a. Metastability and hysteresis are usually characteristic features of first order transitions. The quantitative agreement between MCM and MF is not good. Furthermore, the correct stability of two branches and the corresponding phase transitions are missing from the MF theory.

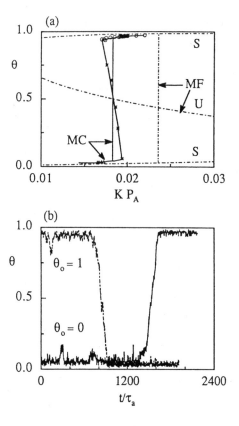

Figure 3. Panel (a) shows the MF isotherm and the MC data, along with their corresponding tie lines. Metastable states are indicated in panel (b) which shows typical reactant coverage time series, starting from a clean ($\theta_o = 0, KP_A = 0.0195$) or a totally covered surface ($\theta_o = 1, KP = 0.0172$). The parameters used are $w/kT = 2$ and $k_R = 0$.

4.3 Cusp points: determination and dynamics. The effect of the value of the reaction rate constant (bifurcation parameter) on the multiplicity of solutions is shown for both models in the bifurcation diagrams figure 4a and 4b. Since first order kinetics are assumed, the reaction of individual atoms is independent of the local structure of the surface. Therefore, removal of atoms from the dense parts of the matrix (clusters) takes also place. This results in destruction of clusters which are necessary for the phase transition to occur. Consequently, the bistability regime is reduced. Above a certain value of the parameter k_R/k_d (dependent on the bifurcation parameter w/kT), a single solution is obtained as shown in figures 4a and 4b.

The cusp point (singularity of codimension 2) at which the two turning points coincide can be determined rigorously in MF theory by application of Eq. (11), as it is shown in the inset panel of figure 4b. Due to the lack of a rigorous condition in the stochastic model, two alternative ways are proposed to determine cusp points. The reactant pressure at the pseudospinodal points, $KP_{A,t}$ is first plotted as a function of k_R/k_d in the bifurcation diagram 4b. As the cusp point is reached, the two turning points approach each other, and the coincidence of pseudospinodal points would determine the singular point of codimension two (cusp point). The exact location of this point requires extremely long simulations on large surfaces because exact values of $KP_{A,t}$ are a prerequisite for this calculation.

Secondly, a size effect analysis is used in analogy with the equilibrium Ising model [10]. Figure 4c shows the amplitude of fluctuations in coverage, $\Delta\theta$, as a function of k_R/k_d at constant reactant pressure. The maximum of the fluctuations should indicate the location of the cusp point. It is observed that below a certain reactant pressure, no large fluctuations dominate the dynamics of the system. This probably indicates that the simulations are performed in the bistability regime where a relatively large perturbation brings the system from one state (a low coverage branch) to the other (a high coverage branch). Once the system has left its original state, it does not seem to return to its initial state, at least during the computational time, if its size is large enough.

Very large fluctuations characterize the dynamics of the system when k_R/k_d is slightly greater than the cusp point value, i.e. as the cusp point is reached from the unique solution subspace. For the calculations shown in figure 4c, we used a constant number of surface sweeps of approximately 120000. A typical time series exhibiting fluctuations is shown in figure 4d. The dynamics of the time series obtained by the MC simulations are examined by using the correlation integral method [8,24]. We have found that the correlation exponent is proportional to the embedding dimension as the latter is increased, indicating that the dynamics are dominated by stochastic noise [8].

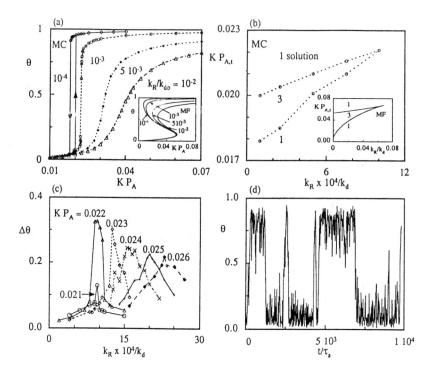

Figure 4. Effect of reaction rate constant, k_R, on metastability for $w/kT = 2$. Panel (a) shows isotherms obtained by the MCM for different values of k_R. Isotherms obtained by the MF for the same parameters are shown in the inset. Panel (b) shows turning points as a function of k_R. No multiple solutions can be found above a certain value of k_R. Notice that the values of k_R obtained by the MCM and the MF model differ almost by two orders of magnitude at the cusp point. Panel (c) depicts the amplitude of fluctuations, $\Delta\theta$, of a 15×15 square lattice as a function of reaction rate constant k_R, for various values of reactant pressure. Very small fluctuations are found below a certain pressure. Panel (d) shows a time series obtained from a 15×15 surface close to the cusp point shown in panel (b) ($KP_A = 0.022$, $k_R/k_d = 9.510^{-4}$).

4.4 Degree of deviations between stochastic and deterministic solutions. The existence of nonlinearities in the governing equations implies a dependence of the corresponding nonlinear rates on the local environment of each atom. Hence, fluctuations occurring in various processes can give rise in the development of spatially nonuniform solutions. In the presence of attractive interactions, surface diffusion does not probably lead to stirring of adatoms which would reduce the spatial inhomogeneities and improve the predictions of MF theory. Figure 2a is an example of this situation.

Removal processes, such as desorption or reaction, when they do not depend on local environment, can partly destroy the formation of clusters if they are sufficiently fast. This is a kind of "stirring" occurring through the coupling with the gas phase. This behavior is observed for both model systems, i.e. unimolecular and bimolecular reactions. When the desorption rate of reactants is sufficiently high, good agreement between deterministic and stochastic model can be achieved, as for example in the AB model system. At sufficiently low desorption rates, randomization of the surface occurs slowly compared to adsorption (slow "stirring") and deviations of the MF model from the MCM are found [11,12]. However, "stirring" in heterogeneous systems through a fast Fickian diffusion or a removal process cannot be experimentally controlled (e.g. by increasing the temperature) without affecting the dynamics of the entire system (all rates), i.e. "mixing" is an intrinsic property of the system which varies with experimental conditions and nature of system.

In the following two sections, the problem of Hopf bifurcation is examined by simulation of Eqs. (3) and (4).

4.5 Limit cycles. Self sustained oscillations can be found in the MC simulations when attractive interactions are sufficiently strong so that multiplicity of the isotherm occurs, as shown in figure 5. Panels 5a and 5b show time series of coverage and reactant pressure predicted by MF and MC simulations, and panels 5c and 5d depict the portrait of the system in phase space along with the isotherm obtained from constant pressure simulations.

The dynamics of the time series were first examined by using the correlation integral method which is used to distinguish random time series from time series which stem from deterministic chaos [8,11,24]. The correlation integral was calculated from the pressure time series obtained from a 30×30 surface which consisted of 10000 points separated by about 33 Monte Carlo steps. Below a certain length, the slope of the curves is constant, ~ 0.95, indicating a high temporal uniformity of the reactant pressure. The Fourier spectrum obtained from the pressure or coverage time series indicates a dominant characteristic frequency which is the inverse of the mean period of the oscillations.

The amplitude and frequency of oscillations remains nearly constant as the size of simulation array increases from 10×10 to 40×40. The simulations are performed far from the cusp point where the dynamics are characterized by large and irregular fluctuations [8]. The coverage and reaction rate oscillations are driven by phase transitions due to absorbate-absorbate attractive interactions and are synchronized by oscillations in the gas pressure.

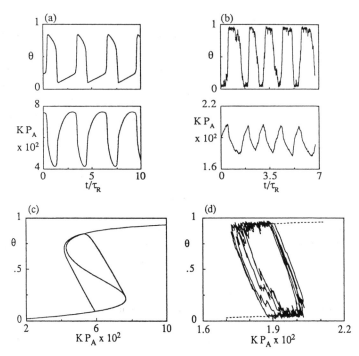

Figure 5. Panels (a) and (b) show time series of coverage and reactant pressure obtained by the MF and the MCM. Trajectories on the $\theta - KP_A$ plane for the MF model and the MCM are shown in panels (c) and (d) respectively. The parameters used in the MF model are $w/kT = 1.5$, $KP_{A,o} = 0.1005$, $KP^* = 0.03$, $k_d\tau_R = 1000$, $k_R/k_d = 0.00333$ and in the MC simulation are $w/kT = 2$, $KP_{A,o} = 0.024$, $KP^* = 0.00144$, $k_d\tau_R = 6.5810^4$, $k_R/k_d = 10^{-4}$, and surface size 30×30.

Since the oscillations are not exactly periodic, the influence of initial conditions on the time evolution of the system is also examined to extract the Lyapunov exponent [25]. Simulations are carried out starting with different initial coverage and surface configurations (adatom distribution), keeping constant the other parameters. With a different initial coverage, the oscillations are again observed, but they start from different points of the limit cycle. The trajectories in phase space are very close to each other for all cases studied. The absolute value of the difference in coverage traces obtained from different values of initial coverage reveals no exponential dependence on time. Recalling the randomness of the stochastic model, the simulations indicate that the oscillations are not strongly affected by the initial conditions.

4.6 Role of defects on periodic solutions. In the previous section, it was demonstrated that mass transport through the gas phase can be a sufficient mechanism for spatial self-organization of the system. This communication is believed to be the dominant mechanism of synchronization on $Pt(110)$ surfaces [26]. Clusters formed on the surface by fluctuations are nucleation centers for subsequent growth and oscillations. Similar nucleation centers exist on real surfaces close to imperfections. Hence, interactions between defects are examined as a possible source of synchronization through the reaction-diffusion phase, i.e. the surface [14]. Reaction-diffusion coupling across the surface is probably the dominant mechanism for oscillations on $Pt(100)$ surfaces [26]. Waves emanating from different locations of the surface (probably defect sites) interact resulting in non regular macroscopic oscillations [26].

In our model, it is assumed that atoms adsorbed on defect sites are not removed by any annihilation mechanism. These atoms interact with other atoms by attractive forces and since they are not destroyed, they constitute nucleation centers. However, there is no obvious reason why all of these nucleation centers should be synchronized. Study of the influence of defects on the phase transition reveals that defects facilitate nucleation and shift the transition (from a low to a high coverage) to a lower pressure. Therefore, defects reduce the pressure amplitude [8] which synchronizes the oscillations.

Figure 6 shows the effect of several different distributions of defects on the oscillations, for 4% imperfect sites. From panel a to c, the number of defect sites (black squares) which can interact with reactant atoms (dotted squares) increases, i.e. the importance of coupling through the surface becomes stronger. The simulations show that in the presence of defects the oscillations become more chaotic (for the same parameters used for a perfect surface). Single defected sites assist the nucleation and the oscillations vanish for the same defect concentration, as it is shown in figure 6c2,c3. As the oscillations become more irregular, a size dependence is also observed, figure 6b1-c3. This indicates that identical unit cells are randomly phased oscillators, and the reaction rate should be a time invariant over a large number of these oscillators. By decreasing the inlet pressure, oscillations can be still found for small defect concentration, but the amplitude of pressure is small and the oscillations are less regular compared to the oscillations obtained from a perfect surface.

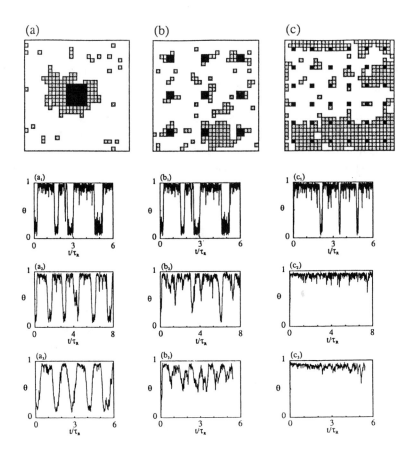

Figure 6. Effect of defect distribution on oscillations for a total defect concentration of 4%. In panel (a) defects are located in one patch at the center of the surface. In panel (b) the unit cell consists of 100 atoms with 4 defect sites at the center. In panel (c) the unit cell (oscillator) contains 25 sites with one defect site. The surface snapshots (a)-(c) are from a 30×30 surface at $t/\tau_R \sim 0.017$. Oscillations in θ are shown in (a_1)-(c_3). On the vertical axis the size effect dependence is indicated. (a_1)-(c_1) are obtained from a 10×10 surface, (a_2)-(c_2) from a 20×20 surface, and (a_3)-(c_3) from a 30×30 surface.

5. Conclusions.

The dynamics of heterogeneous systems have been examined by studying generic model systems with a stochastic method (MCM) and the traditional deterministic approach (MF). The cooperative phenomenon caused by absorbate interactions and the role of fluctuations on the dynamics have been investigated. Spatially nonuniform solutions can characterize the distribution of atoms on a surface whenever nonlinear terms appear in the ODEs. Phase transitions can be frequently associated with these nonlinearities and their role on the stability of the system is crucial. An experimental probe of the local field reveals large fluctuations which stem from stochastic noise (strong correlations) in many cases, especially close to cusp points.

The above features of complicated dynamics which are often observed in experiments are missing in the solution obtained from ODEs. Nonlinearities in the ODEs describing the dynamics of the system imply a dependence of rates on the local vector field and the MC calculations suggest that under such circumstances the MF theory can be only an approximation. The validity of approximation depends on the values of the parameters, the importance of migration, and the surface defects, i.e. the intrinsic "stirring" set in the adlayer. Therefore, utilization of a uniform field to model these systems with ODEs results in deviations from the exact solution of the system and lack of knowledge concerning stability and phase transitions.

On the other hand, even though PDEs can give a spatial variation of the vector field, in the presence of attractive interactions ($w > 0$) or local disturbances of the vector field (defects), they are structurally incorrect. In PDEs of the form (16), surface diffusion tends to destroy the concentration gradients. Therefore, the need for stochastic modeling which provides an exact solution of model systems is very striking. Using the MCM, we have demonstrated the efficiency of the technique to model the complicated dynamics of heterogeneous systems such as multiple solutions and stability, Hopf bifurcations, and the problem of synchronization through surface defects.

REFERENCES

[1] L.F. RAZON AND R.A. SCHMITZ, Catal. Rev. – Sci. Eng., 28 (1986), p. 89.
[2] E. WICKE, P. KUMMANN, W. KEIL, AND J. SCHIEFLER, Ber. Bunsenges. Phys. Chem., 84 (1980), p. 315.
[3] P. MOLLER, K. WETZL, M. EISWIRTH, AND G. ERTL, J. Chem. Phys., 85 (1986), p. 5328.
[4] M. EHSASI, M. MATLOCH, O. FRANK, J.H. BLOCK, K. CHRISTMANN, F.S. RYS, AND W. HIRSCHWALD, J. Chem. Phys., 91 (1989), p. 4949.
[5] I. KEVREKIDIS, L.D. SCHMIDT, AND R. ARIS, Surface Sci., 137 (1984), p. 151.
[6] M.A. McKARNIN, R. ARIS, AND L.D. SCHMIDT, Proc. R. Soc. Lond., A 415 (1988), p. 363.
[7] S. JAKUBITH, H.H. ROTERMUND, W. ENGEL, A. VON OERTZEN, AND G. ERTL, Phys. Rev. Lett., 65 (1990), p. 3013.
[8] D. G. VLACHOS, L.D. SCHMIDT, AND R. ARIS, Surf. Science, (to be published).
[9] R.M. ZIFF, E. GULARI, AND Y. BARSHAD, Phys. Rev. Letters, 56 (1986), p. 2553.
[10] Topics in Current Physics, Monte Carlo Methods in Statistical Physics, ed. K. Binder, (Springer, New York, 1986).
[11] K. FICHTHORN, E. GULARI, AND R. ZIFF, Chem. Eng. Sci., 44 (1989), p. 1403.
[12] K. FICHTHORN, E. GULARI, AND R. ZIFF, Phys. Rev. Lett., 63 (1989), p. 1527.
[13] M. KOLB AND Y. BOUDEVILLE, J. Chem. Phys., 92, (1990), p. 3935 and references therein.
[14] D.G. VLACHOS, L.D. SCHMIDT, AND R. ARIS, J. Chem. Phys., 93 (1990), p. 8306.

[15] See. e.g., A.M. Zhabotinsky, Ber. Bunsenges. Phys. Chem., 84 (1980), p. 303.

[16] P.C. PAWLICKI AND R.A. SCHMITZ, Chem. Eng. Prog., 83 (1987), p. 40.

[17] R.A. SCHMITZ, G.A. D'NETTO, L.F. RAZON, AND J.R. BROWN, NATO ASI Ser., C (1984), 120 (Chemical Instabilities, G. Nicolis and F. Baras (eds.)), p. 33.

[18] R.J. SCHWANKNER, M. EISWIRTH, P. MOLLER, K. WETZL, AND G. ERTL, J., Chem. Phys., 87 (1987), p. 742.

[19] M.E. KORDESCH, W. ENGEL, G. JOHN LAPEYRE, E. ZEITLER, AND A.M. BRADSHAW, Appl. Phys., A 49 (1989), p. 399.

[20] G.A. CORDONIER, F. SCHUTH, AND L.D. SCHMIDT, J. Chem. Phys., 91 (1989), p. 5374.

[21] J.W. CAHN, Acta Metall., 9 (1961), p. 795.

[22] N. METROPOLIS AND S. ULAM, J. Am. Stat. Ass. , 44 (1949), p. 335.

[23] T.L. HILL, An Introduction to Statistical Thermodynamics, (Dover Publications, Inc., New York, 1986).

[24] P. GRASSBERGER AND I. PROCACCIA, Physica, D9 (1983), p. 189.

[25] RUELLE D., Proc. R. Soc. Lond., A 427 (1990), p. 241.

[26] M. EISWIRTH, P. MOLLER, K. WETZL, R. IMBIHL, AND G. ERTL, J. Chem. Phys., 90 (1989), p. 510.